Gardening as a

Jeremy Naydler

Gardening as a Sacred Art

Floris Books

Published in 2011 by Floris Books

© 2011 Jeremy Naydler

Jeremy Naydler has asserted his right under the
Copyright, Designs and Patent Act 1988 to be
identified as the Author of this Work.

British Library CIP Data available
ISBN 978-086315-834-6
Printed in China

Contents

Acknowledgments

The seed of this book was sown in December 1986, when I was invited by my brother Nicholas to talk about the history of gardens and the evolution of consciousness at Cherry Orchards Camphill community in Bristol, where he was a gardener. I shall never forget the warmth and soulfulness of this candle-lit evening, which began with the singing of a beautiful Advent hymn. To him I owe a debt of gratitude for preparing the ground for the ideas that would eventually unfold in the leaves of this book. Since that occasion, my thoughts have frequently returned to the fundamental questions this book attempts to address, for gardening has been my daily employment, pressing me to wrestle with the 'why?' of what engages so much of my time. Out of the many people who have helped me towards deepening these questions, two in particular have been a source of inspiration over many years. One is Patrick Gordon, whose vision of the garden as an outdoor chapel (on good days a cathedral!) sustained and reinforced my sense that it really is possible to work towards the garden becoming a sacred space. The other is Bronac Holden, whose love of the tiniest and humblest plants has ever reminded me that the beauty of the smallest is often greater than that of the showiest of plants.

My decision to at last set down in writing the thoughts that had been gestating for more than twenty years was prompted by invitations to speak at the Prince's School of Traditional Arts, the Gurdjieff Society and the Temenos Academy in London in 2009. These three talks resulted in the booklet, *Gardening as a Sacred Art* (Oxford: Abzu Press, 2009), of which the current book is an enlarged and thoroughly revised version. My thanks go finally to Louanne Richards who, in the long drawn out process of this little book's emergence, has patiently accompanied me on the way, with an amazing willingness to read and re-read the various chapters, both encouraging me and reflecting back the shortcomings and excesses of the prose, which I have done my best to moderate.

Introduction: A Tension Unresolved

For many of us today, it is our gardens that provide our main intimate, daily contact with nature. The garden that we look out upon from a window, the garden that we walk through, stand or sit in, presents us with an immediate contrast with the experience of being indoors, inside the shelter and comfort of our houses. In the garden we see live plants growing, myriad animals – squirrels, birds and insects – moving about, and it is the setting in which we witness the changing moods of the seasons and the weather. Life goes on in our gardens quite independently of us. There we feel the creative power of nature, to which we too owe our existence.

And yet we may also feel that the garden, as well as putting us in touch with nature, is at the same time our own creative project. Many of us are remarkably possessive about our gardens. Indeed, strong emotions – of love and pride or equally, for some of us, despair and shame – accompany our relationship to our gardens. We feel that the garden is not just the expression of nature, but is also *our* responsibility. We are, like it or not, creatively involved in our gardens. We may indulge them or neglect them, we may get inspired by them or feel hopeless about them, we may care tremendously or very little, but whatever our attitude we cannot escape the fact that our gardens present us with a reflection precisely of *our attitude* towards them. The garden is never, therefore, *simply* an expression of nature.

And herein lies a tension that is probably experienced by all gardeners. It is certainly a tension that runs through the history of gardening. The tension is between the extent to which we feel that the garden *belongs to nature*, and the extent to which we feel it *belongs to us* and is an expression of our own creative genius (or lack of it). Just because the garden exists within a polarity, which has on the one side nature and on the other the human ego, historically the attitude of gardeners towards their gardens has either inclined towards *laissez faire* stewardship, or (when taken to its extreme) towards the exercise of total domination.

This has been especially the case from the early modern period to our own post-modern times. The underlying reason for this contrast of attitudes and inclinations is that human consciousness since the Middle Ages has unfolded out of a growing sense of separateness and alienation from nature. Already this was foreshadowed in the Roman period, as we shall see. But before that, the stark division between nature and the human ego did not exist in the way that we experience it now. The garden therefore existed in a different 'space' – a space that was much more readily assimilated to the sacred. Today, we turn to our gardens with a gnawing sense of divorce from nature, albeit with the hope that the garden may offer the possibility of a renewed communion with the natural world. And yet this is not easily achieved, for there always remains the question: a communion on whose terms? The garden can all too often become a battleground between our need to feel that we belong to nature and our desire to be in control.

Today, therefore, two fundamental tendencies can be discerned in the approach people take to gardening. Many people feel that the possession of a garden gives them not only the opportunity but also the right to impress their own designs upon the little bit of nature that has become their responsibility. Such designs will express their own or their family's needs and desires, and thereby they feel the garden's main purpose is fulfilled. For others, however, the garden provides an opportunity to attune to nature, and to follow nature's lead in matters of garden management and aesthetics. If we think in this way, then it is not simply a question of impressing our own designs on nature, but rather of working with the spirit of the place so as to bring it to fuller expression through the decisions that we make.

In the first approach it is assumed that nature should be made to comply with our personal requirements and that it is only right that it be the recipient of our human ideas of order and beauty. But in the second approach there is a reaching toward nature's inherent and 'unconscious' beauty, and a wish to make that the guiding principle. Interestingly, both approaches may result in the feeling that the distance between nature and ourselves is lessened, but in the first approach this is achieved by conforming nature to human aesthetics, whereas in the second approach it is achieved by adjusting our aesthetic choices and decisions to nature.

These contrary gestures are all too plain to see in the history of gardening and, to a greater or lesser degree, they can be detected in most contemporary gardens. They can, furthermore, be experienced by many a contemporary gardener, who feels the pricks of conscience in the decisions he or she makes. In practice, most gardeners are neither wholly despots nor are they wholly benevolent stewards, and there are few gardeners who do not have to wrestle at times with the contrary impulses of imposing order and letting things be. To regard these impulses as living in an irresolvable state of tension, however, would be to miss the crucial fact that the tension itself is symptomatic of the extent to which we feel divorced from the inner life of nature. If we have not yet arrived at a form of gardening that adequately resolves this tension, it is because human beings are still struggling to overcome a deep-seated sense of alienation from the natural world. We are still trying to find ways of closing the gap between nature and ourselves.

Gardening may therefore be seen as an art that has yet to fulfil its true potential. The garden calls to us from the future, where it exists as an as yet unrealised ideal, urging us onwards towards garden forms that, because of the limitations of our imaginative faculty, we have only vague intimations of today. It is from this perspective that we can understand the extraordinary statement of one of the twentieth century's most inspired gardeners:

> The garden hasn't yet happened. In all the existence of great nations that we know of, that is, in our little bit of history, the garden has of course *not yet happened*.[1]

Most of us today have only the dimmest of ideas about how the garden in its most fully realised form might 'happen'. This makes it all the more important that we try to bring what lives within us as an uncertain sense that the garden could be more than it has yet been, into a more conscious formulation of what it could indeed become.

How then can we begin to see our way to resolving the tensions that seem to be intrinsic to the very idea of gardening? And how can we gain a clearer conception of what it is that we should eventually be striving to accomplish in our gardens? First of all we need to look back to the past in order to get our bearings, and then we may be in a better position to formulate something about the direction in which gardening could develop if it is to achieve a fuller realisation. What I hope to show is that if today gardening has come to be conceived as an art, then the way in which we can steer our gardening towards greater maturity is by seeing what it might mean to further conceive, and practise, gardening as a *sacred* art. But to begin with, let us first briefly review the history of European gardening so that, by seeing how we have arrived at the situation we find ourselves in today, we may better orientate ourselves towards the future.

Chapter 1: The Garden in Antiquity

The garden in ancient Egypt

The history of gardens goes back a long way. We know that gardens existed in both ancient Egypt and Mesopotamia. In Mesopotamia the *ziggurats*, which were terraced pyramids surmounted by a shrine, served the religious purpose of linking earth and heaven. They were planted with trees, shrubs and vines from at least the second millennium BC and were probably the original prototype of the legendary 'hanging gardens' of Babylon. Little detailed information has survived, however, and the earliest garden images from Mesopotamia are relief carvings from the much later Assyrian period, which show temples and shrines on hilltops, possibly terraced and planted with various shrubs and trees (Figure 1.1).

Figure 1.1. Assyrian garden surrounding a hilltop temple. Seventh century BC.

The evidence indicates that in Mesopotamia and Egypt gardening was practised in a sacred context, for in both cultures the most important and spectacular gardens belonged to the temples. These temple gardens provided the model for other gardens, which, to a greater or lesser extent, followed the same basic design. Since our knowledge of Egyptian gardens is much fuller than that of Mesopotamian gardens, it is upon Egyptian gardens that this section will focus. An Egyptian temple garden, probably that of Karnak during the Fourteenth Dynasty (eighteenth century BC), is shown in Figure 1.2. It consists of a rectangular pool (fed from a canal) in which lotus grow, surrounded by papyrus, sycomore figs (*ficus sycomorus*) and vines. In this sacred precinct, just in front of the entrance to the temple, King Neferhotep I ritually presents his queen, Meryet-Ra, with a bouquet.

A profundity of conception underlies the apparent simplicity of design of this garden. In ancient Egypt every temple had a sacred lake, or 'divine lake' (*shi-netjer*) as the Egyptians called it. It symbolised the original waters of creation, from out of which all of life originated.[1] The plants both in and surrounding the lake were not so much placed there for decorative effect as for their symbolic meaning, for they evoked the

Figure 1.2. Egyptian temple garden, eighteenth century BC.

paradisal garden of the gods in the time of beginnings, known by the Egyptians as the First Time. The 'garden of the gods' was also known to the Mesopotamians, and is described in the *Epic of Gilgamesh*, which dates back to the third millennium BC.[2] The temple garden, certainly in ancient Egypt and most probably also in ancient Mesopotamia, is best understood as an evocation of a mythic condition in which humans and gods lived in harmony together in a natural world that was instilled with spiritual power and meaning.

The remembrance and recreation of the First Time was of utmost importance for the Egyptians. It was, in a sense, integral to their way of life; it was part of living rightly or living well. This required that one constantly align oneself to the cosmic order through honouring the gods.[3] For the ancient Egyptians, the outwardly observable forms of nature were all endowed with inner qualities that were connected with divine energies. No plant, animal or natural feature was thought of without there being at the same time a consciousness of its original connection to a god or goddess. In the sunlight-saturated atmosphere of the Nile valley, it was this connection with the divine world that the sacred culture, with its temples and sanctuaries, its rituals, prayers and magic, sought to maintain. The Egyptian garden, too, was part of this effort.

For this consciousness, then, which endeavoured always to attune itself to the world of spirit, there were few plants that were not understood to be the vegetative manifestation of a god or goddess. We may perhaps glimpse something of this ancient consciousness in the depiction of the sycomore fig in Figure 1.3. This tree was sacred to the sky goddess Nut, who is here shown in its branches, offering sustenance to a traveller as he journeys through the Otherworld. The very fact that the tree grows both in this world and in the Otherworld sheds further light on the ancient mentality, for which neither garden nor plant was ever conceived to be entirely physical. The 'location' of a garden and equally of any plant spanned both worlds: it was as much spiritual as physical.

Figure 1.4 depicts a garden centred on a sacred lake, which is typically rectangular.[4] It is surrounded on all sides by an abundance of plants – date palms and doum palms, sycomore fig, mandrake and many others – all of which had specific religious connotations. The

date palm, for example, was sacred to the sun god Ra, the doum palm
to Thoth, the sycomore fig (as we have seen) to Nut and the mandrake
to the goddess of love, Hathor. The lotus, a little hard to see growing
out of the water, also had important symbolic associations. According
to myth, the sun god himself was born from a lotus bud, and thus it was
for the Egyptians a potent symbol of spiritual rebirth and regeneration.[5]

Figure 1.3. The goddess Nut appears in her sacred tree, the sycomore fig.

The owner of the garden, therefore, who stands on a boat in the midst of the lake, is immersed in a religious environment. The depiction most likely shows him as engaged in a ritual act rather than merely enjoying a pleasure-ride, for he is emulating the sun god Ra at the beginning of creation, who similarly appears in the primeval cosmos on board his sun boat. Such a garden quite possibly does not even correspond to an actual physical garden, but rather represents an imaginative archetype of an ideal that every ancient Egyptian garden attempted to capture: it was intended to evoke the First Time, when the bounteous world of nature first became manifest, and when the heavenly gods walked on earth.

If these mythical considerations lead us to think that the ancient Egyptian garden was entangled in a complex web of religious belief and superstition, this would be a peculiarly modern way of misunderstanding the ancient consciousness. For it is important to remember that for the

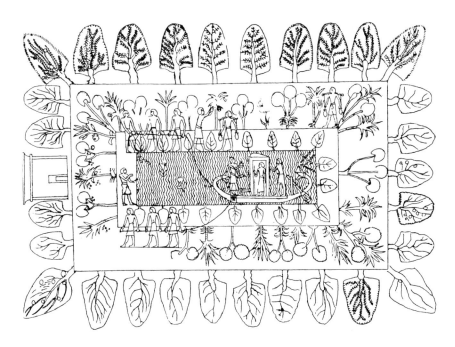

Figure 1.4. Archetypal Egyptian garden, with an abundance of plants surrounding a sacred lake. Fifteenth century BC.

ancient Egyptian nature was not objectified as it is today, but was perceived rather as ensouled, as permeated by invisible beings and agencies, and thus imbued with an inner vitality to which the people of those times were highly sensitive. The ancient consciousness responded intuitively to the spiritual qualities that it discerned in nature. It is therefore less a question of belief and superstition than of a mentality acutely aware of the divine energies that pervaded the natural world, and also acutely mindful of the fact that this awareness was ever endangered and could very easily be lost.

Unless we bear all this in mind, we fail to grasp how crucial it was for the ancient Egyptians that their gardens, with their pool of flowering lotus surrounded by trees, shrubs and flowers, functioned as an evocation of the First Time or paradisal state at the beginning of creation. As such, it was an invitation to specific divine or spiritually charged energies to become present. This is why, referring to the garden that she created at her temple in Deir el Bahri, Queen Hatshepsut stated, not that it was made for humans to enjoy, but that it was made *for the god Amun*, so that he could 'walk abroad in it'.[6]

The same motive lay behind the creation of private gardens. The garden was planted and laid out with an awareness that certain gods and certain spiritual energies would be attracted into it by its design and by the choice of plants. The garden of the noble, Nebamun, (Figure 1.5) is typical in its inclusion of such vital ingredients as the central, lotus-filled pool, and plants such as the mandrake, date palm and sycamore fig around its edges. Nebamun's garden is like a prayer or a magical incantation. To understand it rightly, we must see that for the ancient Egyptian the garden was *a place of communion between the visible and invisible worlds*.

This point is beautifully illustrated in the *Papyrus of Nakht* (Figure 1.6), in which Nakht and his wife Tjuiu come into the divine presence of Isis and Osiris in the garden. We notice its similarity to Nebamun's garden, with its central pool and the familiar plants growing around the water's edge, save for the luxuriant vine that grows, as if drawn by an irresistible attraction, towards Osiris. Whose garden, we might ask, is this? Is it Nakht and Tjuiu's garden or is it the garden of Isis and Osiris? And where is this garden located? Is it in this world or in the Otherworld? Such questions cannot be categorically answered for

Figure 1.5. The Garden of Nebamun, c.1400 BC.

Figure 1.6. The human and divine worlds meet in the sacred garden. Papyrus of Nakht. Late fourteenth century BC.

in the Egyptian context they are false alternatives. The garden created by human beings was always a garden of the gods, a place of epiphany. And the garden of the gods was never so far removed from the human world that human beings lost sight of it altogether. The garden was truly where the human and divine worlds met.

The garden in ancient Greece

If we turn to ancient Greece in the pre-classical period, when the works of Homer were being written (about the seventh century BC), the earlier civilisations of Egypt and Mesopotamia, though hoary with age, were still largely intact. And so in Greece, too, we find that little distinction was made between outer natural phenomena and the world of gods and spirits. Just as in Egypt, in ancient Greece too the sun, the winds and the atmosphere were all gods. So also was every river a god. When in Homer's *Iliad* Zeus called a council of the gods, we learn that 'there was no river who was not there'. The council also included 'the nymphs who live in the lovely groves, and the springs of rivers and grass of the meadows'.[7] For the ancient Greeks, nature was animated by the spirits of the mountains and valleys, forests, lakes, meadows and streams, who were led in their rhythmic dance through the seasons by the god Pan (Figure 1.7).[8] Every plant had its indwelling spirit – the *anthousai* dwelt in flowers, the *dryades* or *hamadryades* in trees. If a plant became sickly, the Greeks understood that it was because the spirit within it had sickened. And when a plant died, what was really occurring was the departure of this spirit (or nymph) from it.[9]

In the ancient world, natural phenomena were rarely perceived as 'objects' – they were normally subjects, with inner qualities and emotions just like those of human beings. Nature was, as Martin Buber put it, experienced in antiquity more as a 'Thou' than as an 'it'.[10] For this reason, the relationship of people in antiquity to the natural world was also a relationship to the community of spiritual beings that animated all natural phenomena. Nature had a rich inner life. In Figure 1.8 we catch sight of two satyrs with a maenad, dancing wildly amongst trees. These beings, part human, part divine, are the Dionysian counterpart of the entourage of Pan. They are perceived in a state of consciousness

Figure 1.7. Two Pan figures, each holding goats, while above three nymphs dance, c.520 BC.

in which the soul-life of the human being shares intimately in the soul-life of nature.

The gods and spirits were as much a part of the world that human beings inhabited as were the rivers, mountains, plants and animals encountered in an outwardly perceptible form. And because humans too are spiritual beings, relationship to this inner world of gods and spirits was at the same time a relationship to a realm to which humans felt that they also inwardly belonged. People felt that both their deeper identity and their spiritual fulfilment were inextricably bound up with putting themselves into a right relationship with the divine world, through ritual, divination, and sacrifice.

The Greeks did not make gardens like the Egyptian temple gardens. Temples were sited rather in locations that were already deemed sacred to the deity honoured, and thus with minimum disturbance of the natural surroundings that were experienced as suffused with the divine

Figure 1.8. Two satyrs and a maenad dance wildly amongst trees. Fifth century BC.

presence.[11] Areas designated sacred, which often subsequently became
the site of a temple, were usually groves of trees dedicated to, and set
apart for, a particular deity.[12] To enter such a sacred grove, which was
deliberately left wild, was to enter a space that was no more meant simply
for human recreation than was an Egyptian temple garden. In this
respect the sacred grove of the Greeks was the opposite of the modern
garden: it was a space set aside for the gods, rather than for humans.

   The nearest the Greeks came to what we would recognise as a
garden was the purely utilitarian farm-garden attested from the time of
Homer onwards, with its vegetable plot, vines, olives and fruit-bearing
trees.[13] This, it would seem, was common enough, but it falls far short
of the more realised concept of the sacred garden of the earlier Near
Eastern cultures, just as it also falls short of the highly designed private
garden that was cherished by the Romans. One reason for this lack of
development of the garden in ancient Greece was that the countryside
naturally had, as still it has, so many beautiful wild flowers, that to have
artificially moved the flowers nearer to their houses would have seemed
quite pointless, if not sacrilegious, to the ancient Greek.[14]

   In time, with the growing concentration of the population in
cities, public parks within or just outside the city became increasingly

common. These were nearly always attached to sanctuaries dedicated to a semi-mythical hero, and were used primarily for gymnastic exercises, and hence were referred to as 'gymnasiums'. Because they were usually planted with trees, which provided areas of shade where people could sit out of the sun, they were also frequented by those who were simply looking for somewhere quiet to go, away from the bustle of the city. Plato used to teach in one of the most famous of the Athenian gymnasiums, the Academy, named after the hero Academus. It soon became customary for every philosophical school to base itself in the midst of a similar kind of simple plantation, perhaps because it was reminiscent of the traditional sacred grove. After Plato's Academy moved to its own private estate, presumably to ensure greater seclusion, the schools tended to occupy private rather than public spaces. It is, however, unlikely that these so-called 'philosophers' gardens' consisted of anything more than an open space planted with a few trees, and they could also contain substantial buildings.[15] Figure 1.9 is one of the earliest surviving representations of such a garden of philosophers, and is thought to show Platos's Academy. Plato has been identified as the third figure from the left, depicted reclining against the tree. The sacred gateway with vases (on the left) and the votive column (centre) are suggestive of the subtle transformation of the original sanctuary of the semi-divine hero Academus into a place where ordinary mortals could meet to engage in a form of contemplative discourse that would ideally reach divine heights.

The Roman garden

It is significant that the depiction of the philosophers' garden in Figure 1.9 is a Roman mosaic from the first century BC, during the period of the Late Republic, for it was precisely during this period that the Romans began to conceive of their gardens as having possibilities beyond the utilitarian function of growing vegetables and fruit trees. Cicero, who knew Athens well, led the way in re-imagining the garden of a Roman villa as a Greek gymnasium, not in the sense of a place for athletic training but rather as a place where one could stroll along promenades and under the shade of trees immersed in philosophical discourse.[16]

Thus an important step in the secularisation of the garden was taken by the Romans, for, once it was transplanted onto Italian soil, the Greek gymnasium finally lost its original connection to sacred ground dedicated to a god or a hero. It became, rather, a setting for human beings to meet each other and talk in convivial surroundings.

By the time of the Late Republic (*c*.150 BC–27 BC), the Roman world was already one in which the gods and nature spirits had to a large extent withdrawn from direct human awareness. Philosophers had begun to write treatises on the nature of the gods from the third century BC onwards, in an attempt to give a rational explanation for

Figure 1.9. Garden of the Philosophers, thought to represent Plato's Academy. Mosaic. First century BC.

what had become more a matter of speculation and belief than living experience. Cicero himself had written just such a treatise, entitled *On the Nature of the Gods*, whose aim was to demonstrate that the world of nature and the world of the divine belong to separate spheres.[17] Stoic philosophy, influential in Rome from the late second century BC onwards, had a predominantly materialistic view of the natural world, as did Epicureanism, which elaborated a thoroughgoing atomistic and mechanistic philosophy of nature. While the gods of popular religion were at least accommodated within the Stoic worldview, they were totally marginalised by the Epicureans. That these two philosophies should have sustained a position of pre-eminence for many centuries is indicative of the growing detachment of the educated Roman from the inner life of nature.

Despite this fact – or perhaps because of it – there nevertheless existed a degree of nostalgia for a previous age, when the spirits of the forest, spring, cave and meadow were still present to human consciousness. This nostalgia is expressed in poets such as Horace and Virgil during the Late Republic, and subsequently during the Early Empire (27 BC–AD 180) we find it in Ovid, and (though not a poet) Pliny the Elder.[18] But it was not possible for the educated Roman to recapture or return to the older clairvoyance. Their relationship to nature had changed. Plutarch, writing in the Early Empire period, during the reign of the emperor Trajan, grasped this fact, explaining the widespread decline of the oracles as being due to the withdrawal of the spirits or *daimones* from human awareness.[19] He also recounted the deeply shocking story of the death of Pan, who, of all the gods, was most closely identified with the life-forces of nature. Plutarch tells how, almost one hundred years previously (during the reign of the emperor Tiberius), a certain Greek ship was passing near the coast off the island of Paxi, and a ghostly voice was heard to call out: 'When you come opposite Palodes, tell them that Great Pan is dead.' In awe and trepidation the crew sailed on, but as soon as they came opposite Palodes, the winds mysteriously died down and the ship was becalmed. Then the pilot of the ship, Thamus, took courage and shouted: 'Great Pan is dead!' Even before the words had left his lips, the sound of dreadful wailing and lamentation arose from myriad voices of invisible beings on the shore. According to Plutarch, this story soon circulated in Rome, and the pilot was called to give an account of it before the emperor Tiberius, who was convinced of its truth.[20]

The sense that the world of spirit was retreating before the advance of rational consciousness was accompanied by the growing feeling of detachment from nature, which was increasingly seen as something 'out there' to be viewed, acted upon and exploited by human ingenuity and technological power. In the Early Empire, this detachment from nature is well exemplified in the paintings of panoramic scenes that appear on the walls of various Roman villas. Their skilful use of the principles of perspective suggests that nature was experienced less as a numinous presence to which the human being must relate as subject to subject, than as an objectified landscape with its various features spread out before the gaze of the human onlooker.[21]

In Figure 1.10, we see one such naturalistic wall painting from the Early Empire period, in which nature is depicted in a strikingly 'modern' way. It shows a flower-filled orchard teaming with birds, tailing off into scrubland, with the blue background giving the impression that the scrubland softly recedes into the distance. Whereas in an equivalent ancient Egyptian image, every feature would have had a symbolic meaning, we may be fairly certain that here none of the plants or birds depicted was intended to carry a specific symbolic resonance. In this fresco, nature is shown without any underlying spiritual content or context.[22] The consciousness that could produce such an image of an objectified nature, as something to be carefully observed and impressionistically portrayed, was an unavoidable factor that now asserted itself in human relations with the natural world. This consciousness inevitably determined the kind of gardens that were created by the Romans.

From the first century BC onwards, the Romans came increasingly to value panoramic views. Seneca, writing during the first century AD, commented on the fashion to build villas beside the sea or on hilltops in order to command a good view.[23] Lakeside properties were also favourites. The ideal view, however, was not onto wild nature, but onto a nature that had been thoroughly tamed. One of the requirements of the Roman garden, especially that of the country estate, was that it should be visible from the house. On the larger estates, landscapes were structurally altered and effectively redesigned in order both to

Figure 1.10. Wall painting from the Villa of Livia, Primaporta, Rome. First century BC.

enhance the view from the villa and to subsume the landscape under the jurisdiction of its human owner. Hills were levelled, avenues of trees were planted in exact straight lines, and paths and beds were set out so as to conform to the Romans' love of order and symmetry. The aim, as the poet Statius pointed out, in his description of the villa and garden of his friend Pollius Felix, was to 'tame' wild nature by using human engineering and design-skill. He says that where nature:

> ... was wild and unlovely, now it is a pleasure to go... Here, where you now see level ground, was a hill ... where now tall groves appear, there was once not even soil: its owner has tamed the place ... each window [of the villa] commands a different landscape.[24]

Notice the association of 'wild' with 'unlovely'. Here the concept of beauty is linked with the concept of the human 'taming' of nature. The experience of beauty in nature, it seems, could not be assured unless she was tamed, and brought under human aesthetic control. It is only fitting, therefore, that in the grounds of the villa of Pollius Felix a temple was constructed, dedicated to the hero who above all others represented human physical prowess, Hercules.[25]

In the Roman garden, we see the introduction of a great many human constructions, from fences, fountains and pergolas to terraces, sunken lawns, ornamental pools, large (and to modern taste often oversized) sculptures, artificial caves and grottoes. In Figure 1.11, a fairly typical design of a highly symmetrical garden is shown, with a central pool and fountain, surrounded by arbours, fenced enclosures and pergolas. The whole garden speaks of orderliness and a structure that is more architectural than natural. We notice that the pool is square-shaped, and rather than living plants for hedging, fences are used to divide the garden up into distinct areas. A firm hand is used here to contain nature, and to bring it securely within the human domain.

Something rather similar is depicted in Figure 1.12, which shows another garden design with symmetry determining the layout of pergolas, urns, trees and statuary, and the use of fencing to delineate the different areas of the garden.

One of the obligatory components of the garden, whether large or small, rustic or urban, was a decorative shrine-like water feature,

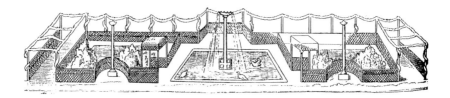

Figure 1.11. Drawing of a fresco, showing an ornamental pool surrounded by arbours, fenced enclosures and pergolas arranged symmetrically.

Figure 1.12. Drawing of a fresco, in which all the garden elements are carefully arranged within a symmetrical design.

containing a fountain, called a *nymphaeum* (Figure 1.13). This feature originated in the ancient worship of the spirits of springs, wherever they bubbled up from the ground, and shrines dedicated to these spirits were revered all over Italy by country folk. In the Roman garden, however, the *nymphaeum* became an essential feature, whether or not there was an actual spring present. In the larger gardens, in order to house the *nymphaeum* a rustic grotto would often be artificially constructed. The *nymphaeum* itself would typically consist of a large basin, often highly decorated, into which water was conveyed and from out of which it flowed, thanks to the Roman mastery of the techniques of channelling water. The *nymphaeum* thus became an ornament to be appreciated as much for its engineering and decorative merits as for its lingering religious function.[26]

The fate of the *nymphaeum* was shared by other rustic shrines (*aediculae*) that originated in the old folk religion. Once introduced into the formal garden setting, they became as much ornamental as religious features, which meant that to a greater or lesser extent they became incorporated within a human aesthetic. This is not to say that genuine religious feelings were absent from the Roman garden, but only that from around the time of the Late Republic onwards the religious consciousness, and the religious forms through which it lived, underwent a kind of fossilisation. Just as Roman religious piety emphasised the outer performance of traditional rites, so in the Roman garden what was originally a focal point of religious feeling increasingly became merely decorative. What this signified was that the inner life was malnourished, and this explains how the eastern mystery cults that grew up in the first century BC became so popular, for they satisfied a craving for religious experiences that the traditional religion was no longer able to provide.[27]

Of particular significance in the development of the Roman garden was the advent of topiary in the Early Empire. The art of topiary, in which plants are sculpted into shapes determined by their owner, was the Romans' peculiar contribution to gardening. It arose out of a consciousness of plants that had become insensible of their indwelling spirits, the *dryades, hamadryades* and *anthousai* of the Greeks. Pliny the Elder traced the origins of topiary to a friend of the Emperor Augustus. By the time he was writing (during the middle of the first century AD),

Figure 1.13. A grotto, with a nymphaeum at its entrance, and a pergola above.

topiary had become a fine art, practised especially on the cypress, myrtle and the box. A skilled gardener was able to shape these plants not only into geometrical figures, but also, so he tells us, into extravagant hunting scenes, or images of a fleet of ships.[28] No doubt in most gardens the less ambitious geometrical shaping of plants was more commonly to be seen. In Figure 1.14, we catch a flavour of the Roman practice, even though the engraving is of topiary in an eighteenth century Viennese garden. While this garden belongs to the early modern era, it nevertheless gives us an impression both of the older Roman art that it is emulating, and the mentality that underpinned it.

Pliny's nephew, the younger Pliny, tells us that one of the skills of his *topiarius* (the slave entrusted with topiary) was to cut box into letters spelling the name of its owner – that is to say, Pliny's own name.[29] Here we see expressed the concept of human ownership of plants, a concept that had no place in the relationship to nature that prevailed in earlier times. It confirms the extent to which, for the Romans, the gods and spirits had withdrawn from nature. The balance of power between human beings and the natural world had dramatically shifted. In a culture based on slavery (the younger Pliny owned five hundred slaves), the master-slave relationship is, in the Roman garden, extended to human dealings with the natural world. The art of topiary is a powerful expression of this, for it involves the eradication of the outer gesture of a

Figure 1.14. The legacy of Rome: topiary in an eighteenth century Viennese garden.

plant, by which we recognise it. The plant is made 'faceless'. In ancient Greek, the word 'faceless' (*aprosopos*) denoted the slave.[30] And so we find the Romans treating plants like slaves, for they are now obliged to conform themselves to the will of their human owner, and bear the impress of an alien shape.

Despite their continuing observance of traditional religious rituals honouring the natural world, and despite the presence of shrines and statues of gods and heroes in their gardens, it is difficult to avoid feeling that what the garden meant to the Roman was above all an opportunity for human beings to celebrate their ability to dominate nature and bend it to the service of human ends. The Roman garden was a statement of human autonomy from the invisible powers within nature that so commanded people's loyalty in previous ages. The garden, as it emerged in the Late Republic and Early Empire, was essentially a creation *by humans for humans*. It was not really a place to commune with the gods, who were made subsidiary to its primary purpose, which was to promote human wellbeing and pleasure, and to reinforce a sense of human power *vis à vis* the natural world.

Chapter 2: The Garden in the Middle Ages

The paradise garden

With the fall of the Roman Empire, gardening collapsed throughout Europe into, for the most part, the merely utilitarian growing of vegetables and herbs. In the Islamic world, however, a garden style developed that, like the Roman garden, was to a large extent dominated by architectural elements but, unlike the Roman garden, was at its core religious in purpose and meaning.[1] The roots of the Islamic garden were in the walled gardens and hunting parks (*pairidaezas*) of ancient Persia. These were protected areas, often of large extent, that were set apart from the surrounding environment, and it was this idea of a walled and secluded area, conceived as a sacred precinct, that was taken up within Islam as an image of Paradise.[2]

The Islamic garden, as it developed from the ninth century onwards, was very formal, with order and symmetry guiding the choice and placement of its various features – the central pool, basin or fountain, the water channels (ideally four of them), and such trees as the olive, date, fig and pomegranate.[3] But the imposition of order on nature was less for the sake of asserting human dominance than for making manifest transcendent sacred principles. In the Islamic garden, both the principles of sacred geometry and the language of religious symbolism were the key determinants of the design. The central basin or pool of water, usually octagonal or rectangular, was – as in the ancient Egyptian garden – the crucial element, for it represented the waters of the Spirit. As such it symbolised the source of rejuvenation and spiritual sustenance at the centre of Paradise. Each of the other features of the garden expressed a profound religious meaning, for their purpose was to make the garden into a reflection on earth of the archetypal reality of Paradise. For example, the four water channels, which divided the garden into four quarters, represented the four rivers of Paradise as described in the *Quran*; and the four kinds of tree represented the different levels of spiritual enlightenment.[4]

The garden, then, was typically composed around the central water feature; and its various components, such as flowerbeds, trees, paths, and areas for sitting were arranged within a geometrical design, breathing harmony and order (Figure2.1).[5] Even when, as in the courtyards of city dwellings, the garden was reduced to its bare minimum of a small centrally placed basin or fountain and perhaps a few plants in pots, its underlying purpose as a sanctuary was always paramount. Indeed, its principal feature was less the trees, plants and flowers, which were quite often omitted altogether because of lack of space, than the water, for this symbolised the spiritual source. Figure 2.1, however, represents the ideal, with the central pool, and its four water channels, surrounded by blossoming almond trees and sunken flowerbeds, with cypress and plane trees in one corner, all enclosed by a high wall.

The purpose, then, of the Islamic garden, was to be a place of prayer and contemplation. In the peace of the garden one came to one's own spiritual centre. This is why the garden not only needed to be secluded from the world, either by being surrounded by a high wall or by being located within a courtyard, but also needed to be designed in accordance with precise geometrical and symbolic principles. If we who, living in more northern climes and loving more natural, informal gardens find ourselves balking at the often rather rigid formality of the Islamic garden style, we may also come to appreciate that this style is wholly redeemed by the underlying intent, which was to create an environment in which the human soul could experience the peace, order and harmony that are characteristic of Paradise. In much Islamic poetry, it is precisely this kind of formal garden that represents the enlightened heart, or 'the garden of the heart'.[6] That is to say, the Islamic garden was not – how could it ever be? – wholly 'out there' for it existed also as an inner ideal, as an aspiration of the soul.

The Paradise symbolism of the Islamic garden was strongly echoed in the medieval European *hortus conclusus*, or 'enclosed garden', which became popular during the twelfth century. It too was meant primarily as a place of spiritual contemplation, and this was because its features,

Figure 2.1. Typical Islamic garden.

which often included a central basin or fountain, along with various trees and flowers imbued with symbolic meaning, and the whole enclosed by a wattle fence or wall, were (as in the Islamic garden) intended to recall Paradise to those who stepped inside it. In other words, the garden was intended to have an effect on the soul of those who entered it and spent time in it, so that they had not just an experience of flower, tree and fountain, but rather that this provoked in them an inner experience of something far greater, so that the veil between the sensory world and the spiritual world was here, in this special place, drawn aside.

This drawing aside of the veil between the sensory and spiritual orders of existence is shown in Figure 2.2, in which God introduces Adam and Eve to the Garden of Eden. They walk out from the portico of a heavenly cathedral into a verdant garden, in the centre of which is a fountain. Either side of the fountain are the two trees of the Garden of Eden, under which a variety of flowers grow and certain animals can be glimpsed. Just as with the Islamic Paradise garden, this garden is surrounded by a wall, from which the four paradisal rivers gush out. What we are viewing here is an archetypal image that belongs as much to the interior world of prayer and contemplation as it does to any outwardly existing garden, for it represents both the heavenly condition from which humanity has fallen, and the promise of the restoration of the soul to God.

This promise of the restoration of fallen humanity to God was contained pre-eminently in the figure of the Virgin Mary, who represented the purified human soul. Because her will was wholly aligned to the will of God, she was able to give birth to Christ, the divine-human being. The *hortus conclusus* was really Mary's garden: in the *hortus conclusus*, one entered the soul-world of Mary. What one experienced outwardly was intended to bring one into the presence of an interior being who called one back to one's inmost self. In Figure 2.3 we see another representation of the garden of Paradise, now with an angel guarding its door. But within it Adam and Eve have been replaced by the figure of the Virgin Mary, while outside stands the beautiful figure of Christ, waiting to be let into the garden. The enclosed garden, then, represents the human soul as bride of Christ. The explanatory scrolls in the image are all quotations from *The Song of Songs*, which are here to be understood as the words of Christ addressed to Mary, now conceived as his bride:

Figure 2.2. The Garden of Eden as the archetype on which the medieval hortus conclusus was based.

Figure 2.3. The 'Mary Garden' as a hortus conclusus on the model of the Paradise garden.

> A garden enclosed is my sister, my bride, a garden locked, a fountain sealed ... a garden fountain, a well of living water, which flows in streams from Lebanon ... Arise you North Wind, and come you South Wind, blow on my garden and let its fragrance flow.[7]

The medieval view of nature

Before pursuing the significance of the *hortus conclusus* any further, we must pause to consider the way in which people in the Middle Ages related to nature; for this bears directly on how we should understand what the *hortus conclusus* meant to the medieval mind. In the Middle Ages, the type of objectivising consciousness of nature that had so strongly been emerging in Roman times was to some extent checked. Most people still felt nature to be alive, ensouled and full of spirits. Several English medieval chroniclers refer to, and indeed speak quite freely of, fairies, elves and other nature spirits.[8] At the same time, because of the influence of the fifth century encyclopaedist, Martianus Capella, who transmitted much of the wisdom of antiquity, and whose work was widely studied in the early Middle Ages, a degree of academic respectability adhered to the view of nature as being populated by a host of invisible beings.[9]

Despite the theological constraints of the time, thinkers of a Platonic disposition, such as Bernardus Silvestris, openly promulgated the view that 'wherever earth is most delightful, rejoicing in green hill, flowery mountainside, and river, or clothed in woodland greenery, there Silvans, Pans and Nerei, who know only innocence, draw out the term of their long life.'[10] What this means is that the medieval experience of nature was not simply of external, sense-perceptible phenomena but also of the inner qualities, energies, moods and atmospheres associated with these nature-spirits. Thus the tendency towards objectification was for a while held back.

Of equal importance in the Middle Ages was the idea of nature herself, conceived as a nurturing feminine figure, and personified as the goddess Natura. In Alain de Lille's *De Planctu Naturae* ('The Complaint of Nature'), written towards the end of the twelfth century, the visible world is described as adorning Natura like a garment. The zodiacal constellations and the planets are jewels in her crown; the birds, fish, beasts, herbs and trees decorate her robe, mantle, tunic and undergarments; and on her shoes are the flowers.[11] The imagery expresses the idea, which in subsequent centuries was to exert a profound influence, that Natura has two aspects: *Natura naturans* and *Natura naturata*. By the former was meant the invisible being whose creative energies give rise to the visible world of manifest forms; while by the latter was meant the physically manifest products of this creativity that we see all around us in sense-perceptible objects and beings. *Natura naturans* was Nature conceived as spiritual subject, *Natura naturata* was Nature viewed as physical object.[12]

In Figure 2.4, Natura is portrayed as a nurturing goddess, suckling the world (or more accurately, expressing milk from her very full breasts onto the world), which rests on her lap. The image is from a late medieval manuscript of Aristotle's *Physics*, written here in Greek rather than in Latin and so at the time accessible only to relatively few scholars. The Greek word *physis* means 'nature', and Natura appears here, appropriately, on the very first page of the manuscript, for she is the subject of the treatise. What is extraordinary about this image is that, while it forms part of the

Figure 2.4. The goddess Natura sustains the earth with her milk.

collective imagination of the time, here it appears in a scholarly text, with the implication that the educated just as much as the uneducated found this kind of personification of nature acceptable.

Natura is represented again in Figure 2.5, looking rather more matronly, as she receives from her spiritual daughters the different kingdoms of nature. Each daughter is represented as being at a different stage of maturity, from youth to adulthood, and carries towards the figure of Natura a different level of existence: at the far left, mineral essence; then life, characteristic of plant forms; then sensation, characteristic of the animals; and finally reason characteristic of human beings. It was generally understood in medieval times that in order to live a fully human life, we must live in accordance with reason. When exercised contemplatively, reason enables us to mirror in our minds the universal cosmic-divine order, and thereby consciously to reflect the greater macrocosm of which we are a part. Without this conscious act of reflective awareness, Natura's offspring (*Natura naturata*) would remain separated from their source (*Natura naturans*), and Natura would in effect remain unconscious of herself. In finally receiving the human child into her arms, Natura could be said to be returning to herself through human consciousness.

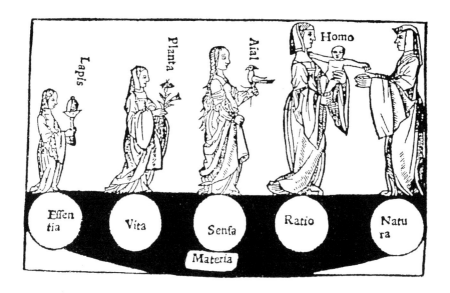

Figure 2.5. The degrees of Nature.

The significance of the enclosed garden

Not all medieval gardens were of the enclosed 'paradise garden' type. By far the most common kind of garden was the 'kitchen garden', neatly laid out for growing vegetables and herbs; orchards, too, containing a range of fruit and nut-bearing trees were also widespread, varying in size according to the wealth of the owner; again, larger pleasure parks with woodland, pastures, ponds and 'rides' were the privilege of the nobility.[13] Many and varied were the gardens of the medieval period, but that which most typifies the spirit of the Middle Ages was undoubtedly the enclosed garden or *hortus conclusus*, which became popular in the twelfth century. For this reason it is to the *hortus conclusus* that we now return, to consider its characteristics further.

From the many portrayals of the enclosed garden that have survived (all from the later Middle Ages), we see that they typically consisted of an informal 'flowery mead' or meadow planted with a variety of sweet smelling herbs and flowers, grown underneath a small number flowering fruit and nut trees, the whole enclosed by a stone wall or wattle fence. Simple turf seats, covered with grass, are also often depicted, near a central basin, pool or fountain (Figure 2.6). Albertus Magnus, writing in the thirteenth century, describes such a garden in some detail in his treatise *De Vegetabilis et Plantis* ('On Vegetables and Plants'). We learn from him that the herbs included rue, sage and basil; the flowers included violet, columbine, lily, rose and iris; and these would have been grown underneath such trees as the pear, apple, pomegranate and walnut. The centrepiece of the garden was a 'clear fountain of water' beside which turf benches, planted with flowers, were placed.[14] His description not only corresponds quite well to Figure 2.6, which dates to the mid-fourteenth century, but is also similar to an early twelfth century description of a garden, contained within *The Life of St Liutgart of Wittigen*. This suggests that the same broad concept of the ideal garden did not change greatly between the twelfth and fourteenth centuries.[15]

Every garden, whether ancient, medieval or modern is the expression of an intention, which may be either dimly or clearly conceived, and also more or less perfectly realised. In the design of gardens, we are considering the deliberate creation of certain effects, and this also applies to the enclosed garden with its flowery mead. What, then, was the intention behind it? The flowery mead was not about domination

Figure 2.6. The poet and composer Guillaume de Machaut sits writing beside a spring, within a walled garden, planted with both trees and flowers. Fourteenth century French manuscript.

or control, but about the enhancement of natural beauty in a direction that was religiously meaningful. As Teresa McLean put it:

> The flowery mead was the *locus amoenus* ['delightful place'] of God's beautiful world, enclosed by man as his own and, insofar as it was possible, improved upon by the addition of more meadow flowers.[16]

Let us take this further. We have seen that, once the enclosed garden was brought into connection with the Virgin Mary, it represented the interior of the soul, transformed and purified, and thus able to reflect the heavenly Paradise of spiritual virtues and archetypes. Conceived in this way, the *hortus conclusus* was not just *associated* with the Virgin Mary, the purest of human beings who, because of her purity, was able to give birth to Christ. It actually *symbolised* her: she *was* the enclosed garden, and the enclosed garden *was* her. This is beautifully portrayed in Figure 2.7, where the enclosed garden is personified in the feminine figure of the Virgin, who seems almost to grow out of the earth alongside the flowering plants.[17]

Figure 2.7. The virgin as the enclosed garden.

In the Middle Ages, the Virgin Mary was by no means simply conceived as a human being. As the sole feminine figure in the religious mythology of Christianity, she inevitably attracted to herself attributes that belonged to the great goddesses of antiquity. Most important of these was the lost goddess of the earth, with

Figure 2.8. Mary as Rose of the World (Rosa Mundi).

whom Mary was unofficially identified through her becoming linked in popular thought to the life forces of the earth and the fertility of crops. In hymns, images and rituals, Mary was consistently identified with the old Mother Goddess.[18] There are many images that associate her with the ripening corn, and in one hymn we read that it is she who makes 'a meadow of delight to blossom'.[19] Mary was more than just the virgin womb that gave birth to Jesus; she was also the womb of creation itself. The symbol of Mary as the *Rosa Mundi*, or 'Rose of the World', powerfully expresses this (Figure 2.8).[20] So while theologically her divine status may not have been recognised any more than that of the goddess Natura, *mythologically* the two feminine figures of Mary and Natura could hardly fail to coincide in the popular imagination.

This means that in the womb-like enclosed garden, with its blossoming 'meadow of delight', one stepped into a sacred precinct. One came into the presence of the divine feminine, the rose of the world, the fountain of life, the *mater generationis* or 'mother of creation' who 'enclosed the whole of heaven and earth within her womb'.[21] This is why the *hortus conclusus* was most frequently depicted with the dominating figure of the Virgin not only centrally placed but also more or less occupying the whole garden.

In Figure 2.9, we see her with the Jesus child, accompanied by angels playing the lute and the harp. Vegetation in the picture is limited to just three flowers (impossible to identify with certainty) and the two trees of Eden. Clearly, this is no ordinary human figure, but a goddess in all but name. To the extent that the experience of being in an enclosed garden involved some sense of being in the presence of Mary, therefore, the purpose of the *hortus conclusus* was far from merely providing a pleasant place to relax and pass the time of day. It served a religious function, and was a space in which one had a heightened awareness of the divine feminine matrix of creation.

Of the many depictions of the Virgin Mary in her garden, remarkably few show the range of plants that might have been grown in it. But one thing is certain, and that is that a significant proportion of the plants would have been perceived as having a quite specific symbolic meaning. This symbolic meaning was not so much imposed upon the plants but rather was seen as belonging to their inner nature, which was expressed in their manner of growth, type of leaf, flower or fruit. The pristine white lily, for example, symbolised the Virgin's purity, the red rose the part Mary played in Christ's passion and her perfect love. The refreshing sweetness of cherries symbolised the joys of heaven, strawberries the fruits of the righteous, while the trefoil leaves of the strawberry symbolised the Holy Trinity. Violets growing low on the

Figure 2.9. The dominant figure of the Virgin Mary in the hortus conclusus.

ground symbolised her humility and the gentle lily-of-the-valley her meekness, and so on.[22] In Figure 2.10, a detail from a late medieval painting, all these plants can be seen.

That flowers should represent moral attributes suggests that for the medieval mind it was not simply that human moral qualities were reflected in plants, but rather that the qualities of plants were capable of being mirrored in the human soul. And in the attempt to accomplish this lay the path by which the human being could become a microcosmic mirror of the spiritual impulses that underlie the created world, by bringing them consciously to manifestation in daily life.[23] To the extent that a person was able to become inwardly innocent, through developing and embodying the Christian virtues, a consonance was

Figure 2.10. A glimpse into the symbolic world of the medieval hortus conclusus.

achieved between what came to live within the human soul and what lived in the innocent world of plants. The *hortus conclusus*, then, was the setting in which the inner life of nature and the inner life of the virtuous human being could begin to reflect each other, to the extent that the human being was able to realise the virtues embodied in the plant kingdom in a human, that is to say, a *moral* way. The garden, with its innocent flowers was therefore like a tutor, whose teaching was to remind people of their inner spiritual tasks.

This is not to say that the *hortus conclusus* was either created or used exclusively for religious purposes, but rather that this religious level of meaning was absolutely integral to it. In the Middle Ages, the secular and the sacred were not separated from each other in the way that they have since become. Living in a secular age, we tend to forget the all-pervasiveness of religious concepts and feelings in a culture such as that of medieval Europe. These determined not just the kind of thoughts people had but also the way in which they perceived and experienced the world, and the way in which they lived their lives.[24] No matter if people fell short of their spiritual and moral ideals (for this will ever be the case), few activities took place without moral and religious thoughts and feelings as their background and matrix. For this reason, no matter what people did in the garden, they would have been aware of the underlying religious symbolism that would have permeated its atmosphere.

It is important also to realise that the Virgin Mary, as well as being – at the level of collective mythology – interchangeable with Lady Natura, was also interchangeable with a human lady as subject of devotion. In the courtly love tradition that blossomed in the twelfth century, reverence for the human beloved was another means by which the lover raised himself to a higher moral level. To the extent that the lover's ardour was transformed by the true Christian love of *caritas*, secular love became sanctified as a route to Paradise. For then it was pure love, in which the knightly lover, conceiving of his beloved in the image of the Virgin Mary, himself strived to become like Christ, just as his beloved strived to become like Mary, now conceived as the bride of Christ. Thus the enclosed garden was ever infused with the warmth of both earthly and heavenly love (Figure 2.11).

Figure 2.11. The hortus conclusus as a garden of love.

Chapter 3: From the Renaissance to the Eighteenth Century

The birth of the perspectival consciousness

The end of the Middle Ages was marked by the gradual loss of the 'symbolic consciousness' in which natural phenomena were experienced as capable of becoming charged with symbolic meaning. Instead, things increasingly came to be seen purely physically, stripped of any connection to the archetypal worlds of the religious imagination. This is not to say that the post-medieval world knew nothing of symbolism. The Renaissance was in many ways far better versed in mythology and symbolism than the Middle Ages, but this had increasingly to be learned rather than being part of the lived experience of nature. People had to educate themselves to think symbolically, whereas in previous ages consciousness was endowed with an untutored symbolic imagination. In the post-medieval age, the human being came to relate to nature more and more as an onlooker, whose inner life was quite separate from anything that belonged to the natural world. Thus it became possible to portray nature divested of any religious meaning in the newly rediscovered art of perspective painting and drawing.

In 1435 Alberti produced his groundbreaking treatise *On Painting*, in which he outlined the principles of perspective painting so that the artist could, step by step, represent the world spatially. In Europe, during the fifteenth century, a tremendous inner struggle took place within the human soul, a struggle in which artists especially were deeply engaged, to re-perceive the world entirely objectively, that is to say in three-dimensional space shorn of any imaginative content and deprived of symbolic undertones.[1] The rediscovery of perspective was one outcome of this struggle. It corresponded to an inner need in people to separate themselves, as human observers, from the objects they observed by interposing spatial distance between subject and object. It was at this historical moment that the modern sense of individuality was born.

In Figure 3.1, we see how the technique of rendering a subject in perspective was practised. The extent to which the woman in the picture undergoes objectification corresponds exactly to the extent to which the artist, in adopting a 'perspectival consciousness', asserts himself as an individual with his own unique standpoint.[2] Just as the woman is objectified by means of the grid of squares, so the artist secures himself in his singular stance towards the world with the aid of the vertical shaft by means of which he steadies his eye.

We wonder who this woman is. Could it be that she is Natura herself? And could it be that this artist, with his cool, concentrated focus, is the precursor of the new scientific consciousness? It will not do simply to say that this image is merely demonstrating the technique of perspective painting. The artist who made this etching, Albrecht Dürer, was still medieval enough to recognise that the reclining feminine form, positioned in front of a landscape with hills and trees, symbolised Natura. And yet he was also modern enough to wish to capture and portray the new phenomenon of the artist as a dispassionate onlooker, whose dedication was to representing objects with complete fidelity to their physical appearance.

As a young man, Dürer had visited Italy, and we see in Figure 3.2 a landscape painting produced by him in 1495, showing his early

Figure 3.1. Training the new perspectival consciousness.

Figure 3.2. North Italian landscape, by Dürer, c.1495.

mastery of perspective. Here nature is portrayed as it appears to the eye of the detached observer, whose interest is wholly in capturing the physical details of the landscape, as seen from his unique vantage point. What we do not see in this picture is any sign of a nymph, or the Virgin Mary; nor in any of the objects represented is there any hint of symbolic connotation. There are no angels, gods or spirits here, and neither is this landscape the setting of a religious scene. This is not to say that Dürer was not a religious man, nor that his vision of nature lacked love. Far from it! But what Dürer excelled at as an artist was physical accuracy, and what he depicts here is a landscape that is simply *a landscape* extended in three-dimensional space.

Thus, in the fifteenth century, the first steps toward the disenchantment of nature are taken. Although she will remain feminine in the collective imagination for a long time to come, in the centuries following the fifteenth she first loses her status as a spiritual subject or goddess and then, as nature is understood increasingly in the mechanistic terms of the new science, she loses her soul and finally her life.[3] Meanwhile, the old view that human beings fulfil themselves by attuning to and reflecting the divine-natural world order is gradually replaced with another view of human fulfilment. It is by developing our own unique human gifts and expressing our individual creative talent, our special 'genius', that we best fulfil ourselves. The concept of humanism emerges simultaneously with the development of perspective, because in both the vitally important starting point is individual ego-consciousness. Rather than seeking to align oneself with the divine, the centre of gravity now shifts decisively from the divine to the human.

The formal garden revived

This new spirit is expressed in a style of gardening that picks up from the point at which the Romans left it. As well as his treatise on perspective, Alberti wrote a treatise on architecture, in which he set forth the principles of villa and garden design, drawing heavily on classical authors.[4] Once again the idea asserted itself in the minds of wealthy landowners that, through the garden, one may demonstrate one's power of ownership of nature by compelling her to serve human designs and to express human aesthetic ideals.

Alberti himself was commissioned to create several gardens, which he conceived as extensions of the house. These were typically laid out geometrically, with paths in straight lines, beds arranged in symmetrical patterns and trees planted in orderly rows. As in the Roman garden, topiary once more became an important feature, so plants that could be clipped or trained were particularly favoured. The topiary was often fantastical. One of Alberti's gardens had topiary in the form of apes, donkeys, a bear, oxen, giants, and a harpy, while another contemporary garden – that of Cosimo de Medici – had topiary in the form of elephants, a ship with sails, a wolf fleeing from dogs and an antlered deer.[5] Along with topiary, the Renaissance garden was embellished with fountains, sculpture and such architectural features as stairways and balustrades. These latter were of crucial significance for they enabled the garden to take the form of a series of terraces, flattening the natural lie of the land and rendering it the passive recipient of the required design. Thus the garden became the province of the architect. And so during the Renaissance the concept of the garden as integral to the house or villa, and hence as a work of architecture to be designed on the drawing board, was established. During the fifteenth and sixteenth centuries this became *the* garden style to be emulated not only throughout Italy but also throughout Europe.

The early Renaissance gardens were on a relatively modest scale, linked to the house by loggias and terraces, with pergolas and covered walkways contributing to the sense of architectural dominion within the garden. An idea of the atmosphere of such a garden can be gleaned

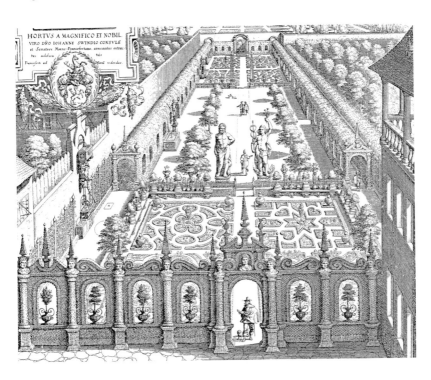

Figure 3.3. Garden in the early Italian Renaissance style.

from Figure 3.3, which, although it depicts an early seventeenth century German garden, reproduces many of the features of the early Italian Renaissance style. We notice at once that every inch of it is designed, and thoroughly subsumed within human aesthetic values, with the two giant-sized statues of Jupiter and Mercury emphasising that these values derive from ancient Rome. The medieval idea of the garden as a sanctuary has given way here to the idea of the garden as a display of the classical culture of its owner.

The later gardens of the Renaissance were on a much grander scale, with large areas architecturally shaped and axially related to the villa to which the garden belonged. In the garden of the Villa d'Este, first set out in the early sixteenth century, the Renaissance garden reached its apogee (Figure 3.4). The Villa d'Este garden was a triumph of the architectural style, with magnificent stairways, terraces, trellis and statuary. It was also a triumph of hydraulic engineering: there were extraordinary water effects – from impressive cascades to fountains that roared or hooted; a wondrous water organ, and jets of water that were artfully contrived to make emblematic shapes. Along with all this

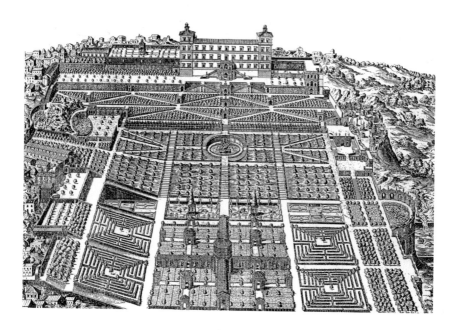

Figure 3.4. The gardens of the Villa d'Este, in the late sixteenth century.

were avenues and covered walkways that led the awed visitor through a series of tightly designed outdoor rooms, in which nothing was left to chance. Given all the stunning effects, both architectural and hydraulic, the plants themselves were the least important element. Marshalled in straight lines or forced to describe elaborate patterns on the ground, they were merely the foil to the more spectacular demonstrations of engineering prowess and technological wizardry.

The garden of the Villa d'Este, in so categorically demonstrating human sovereignty over nature, was meant to glorify the Este family and enhance the renown of its owner, Cardinal Ippolito d'Este. The garden achieved this by locating both the Este family and the cardinal himself in a world of classical myth and symbol. The whole garden was supposed to represent the mythical Garden of the Hesperides, from which Hercules had stolen golden apples, and the sculptures (mostly of Greek or Roman gods, goddesses and heroes), along with the grottoes and water features, were intended to be experienced by visitors as a multi-layered allegory.[6] The famed double terrace of one hundred fountains, for example, was an allegory for Hercules' cleansing of the Augean stable. Hercules was indeed the spiritual patron of the garden, which was a Herculean labour in its own right. The other Greek hero celebrated in the garden was Hippolytus, son of Theseus, with whom Ippolito d'Este felt a special connection because he was his namesake. It is hard to avoid the feeling that, although redolent with classical symbolism, the principal aim of the garden was – like the gardens of so many Renaissance cardinals, kings and potentates – to bolster the self-importance and prestige of its owner by situating him in the centre of a mythic world.

In the course of the seventeenth century the Italian style became very popular throughout Europe. One of the most famous gardens outside Italy was created at Heidelberg for Frederick, the Elector Palatine and his wife Elizabeth, daughter of King James I. The so-called *Hortus Palatinus* was regarded at the time as an eighth wonder of the world, having geometrical designs of great complexity, mythological statues (one of which spoke when the sun's rays struck it), and an abundance of mechanical marvels such as a water organ, and musical fountains hidden in grottoes, creating an altogether fantastical atmosphere.[7] Its deeper purpose, however, was to set out in a veiled form certain esoteric truths, including stages of the alchemical work, and in key respects its design resembled the garden described in the Rosicrucian mystical fantasy, *The Chemical Wedding of Christian Rosencreutz*.[8] Constructed at the height of the Rosicrucian movement, the *Hortus Palatinus* presented the world with a new concept of the garden as a kind of mystery centre, both an encyclopaedia of symbolic knowledge and the setting for mystery dramas and pageants.[9]

The *Hortus palatinus*, however, was also a paradox, a contradiction in terms. Like the Villa d'Este, there was no thought of honouring the spirit of the place, or of bringing about a gentle enhancement of nature towards the paradisal state, such as one finds in the medieval enclosed garden, and which was an ideal at the kernel of Rosicrucian thought.[10] Its designer – the celebrated hydraulics engineer Salomon de Caus – saw fit to use gunpowder to blast away huge swathes of the rocky hillside and fill in the valley below in order to create the flat terracing needed for laying out the geometrical designs of the garden on the required scale. Despite its mystical pretensions, nature was neither honoured nor celebrated in this garden, but was rather subdued and dominated by the engineering and technical skills of its creator (Figure 3.5).

Figure 3.5. The castle and garden at Heidelberg, 1620.

In England the garden at Wilton (Figure 3.6), is another example of a design in which geometrical and emblematic motifs predominated, serving the aim (as at Heidelberg) of making the garden into a theatrical setting for feasts and the performance of masques during the reign of Charles I. The garden required the flattening of an area four thousand feet long and four hundred feet wide, and was divided into three sections: the first consisted of box parterres with fountains and statues representing love and chastity; the second was a strictly measured out 'wild' thicket of trees either side of the river Nadder, with statues of Bacchus and Flora; and the third consisted in concentric walks under cherry trees, around a centrally placed statue of an antique warrior, the Borghese Gladiator. At the end of the garden was a long terrace incorporating a grotto, with the usual amazing hydraulic effects, including mechanical singing nightingales. At Wilton, as at Heidelberg, the intention was less to celebrate the glory of the owner than to affirm metaphysical and moral truths that were the subject of

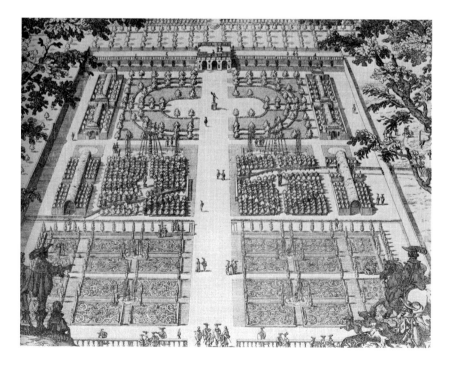

Figure 3.6. The garden at Wilton House, Wiltshire, 1645.

the elaborate masques for which the garden was the stage-set. With the establishment of the so-called 'King's Peace', from 1629 to 1640 when Charles I ruled without Parliament, the Caroline court understood itself in moral terms as purging vice from the land. It was during this period that Wilton was created, and Charles I loved to visit it every Summer, for it subtly symbolised the establishment of the civilising virtues both in its layout, which so clearly imposed order on unruly nature, and in the moral messages conveyed by its statuary, positioned at critical junctures in the progress through the garden.[11]

The images of the Villa d'Este, Heidelberg and Wilton House are typical of engravings of the gardens of this period, in that the view always is from outside and above. This only reinforces our impression of the garden being *imposed from without* upon a reluctant nature. While the drastic engineering works needed to create the Villa d'Este and Heidelberg gardens were not needed at Wilton, there is no sense that the garden emerges from the site itself, reflecting something of the indwelling spirit of the place. Indeed, the river Nadder that runs through the centre of the garden barely interrupts the perfect symmetry of the plan, and its effect on the overall layout is hardly noticeable. One of the characteristics of the Renaissance garden style was that, just like the construction of a building, it required level ground. The very act of levelling the ground could be seen as the signature of the human will asserting itself over nature's unconscious sensuality, in order to replace it with the 'virtuous' configurations of symmetrical design and to construct upon it artificial contrivances.

As well as the usual classical statuary, grottoes and special water effects, Wilton was notable too for its box parterres or 'knot' gardens, shown at the bottom of the engraving. Although parterres were first introduced by the Italians, it was the French who really developed them, from the latter part of the sixteenth century onwards. The word *parterre* simply means 'on the ground', and it signifies an intricate, curvilinear geometrical design best appreciated from a vantage point above it. The attitude that lies behind the development of the parterre epitomises the Renaissance and early modern garden. It is that in themselves the forms of nature have no intrinsic meaning or beauty, or if they do, then for the purpose of making a parterre this is best ignored. Nature simply provides the elements from which human beings will reshape her into patterns that are meaningful and beautiful to the human spectator. Such an attitude arises from the conception of the realm of meaning and value as having no existence outside the human mind. It is *we* who contribute metaphysical significance and beauty to the world, which in itself is regarded as bereft of metaphysical content and aesthetic merit. The assumption is that nature has nothing to give to the garden apart from the raw material that human beings are then privileged to mould into rational designs.

Parterres were typically created using low evergreens (like box) or aromatic herbs (clipped so as better to demarcate the arabesque shapes) and gravel paths or coloured sands to fill the intervening spaces. Figure 3.7 is an example of a seventeenth century French parterre. To view it is at once to marvel at the human ingenuity that has gone into its creation, and to despair at the way the poor plants are forced to form patterns that are so alien to their innate character and inclination. Here we see nature tortured, dismembered and reassembled again by a mentality that has lost all connection with that living, creative and ensouled aspect of nature that the medievals referred to as *Natura naturans*.

Figure 3.7. French parterre, from the seventeenth century.

The assertion of human supremacy over nature in the formal garden style reached its awesome climax in the work of Le Nôtre during the later part of the seventeenth century, when he designed the gardens at Vaux-le-Vicomte for the finance minister of Louis XIV, Nicholas Fouquet, and was subsequently employed by the king himself to create the gardens at Versailles. The overriding purpose of these gardens was to impress visitors with their magnificence. The scale of Vaux can be gauged by the fact that the garden required the destruction of three villages, the diversion of a river and the employment of eighteen thousand labourers.[12] Figure 3.8 gives some impression of the enormity of the project. After clearing and levelling the ground, exquisite parterres, pools with fountains, elegant canals, water cascades and promenades were laid out in accordance with the principles of perspective. These principles were treated from a theoretical point of view by Descartes in his essay on optics, *La Dioptrique* (1637), and were discussed with a view to their practical application in gardening manuals of the time, in particular Boyceau's *Traité du Jardinage* (1638).[13] Whether in conscious imitation of the Romans' deference to Hercules as god of landscape gardening, or whether by a natural

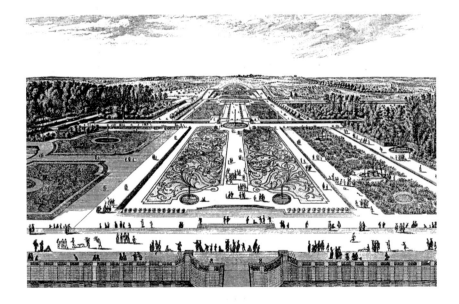

Figure 3.8. View of the garden of Vaux-le-Vicomte from the castle terrace.

movement of the imagination towards the archetypal figure of the mighty hero, a colossal statue of Hercules was erected on top of the hill that framed the garden at one end, and was the culmination of the long promenade that formed the main axis of the garden.

No one had seen anything quite as grand as Vaux, which provoked the envious king to dismiss his minister and commission Le Nôtre to surpass the achievement of Vaux at Versailles. The scale of Versailles far exceeded that of Vaux, needing the labour of no less than thirty-six thousand men and six thousand horses over a period of many years.[14] Versailles was a megalomaniacal exercise in the domination of the natural world by a man who was concerned neither with the ecological limits of the site, nor with any concept of gardening in harmony with nature. The intention was rather to subject the landscape to an arduous mathematical discipline, through regimented plantations, broad promenades, large and elaborate parterres and an excessive number of fountains and pools.[15] The enforcement of order had to be vast in extent, for only thereby could the garden achieve true magnificence. It was Le Nôtre's ability to command and subjugate nature, and so to transform it that every aspect of the garden was an expression of the seemingly unlimited range of the king's power which bestowed upon the garden its real meaning. Both in its layout and in the mythological themes of the statuary, Versailles was a celebration of the rationality of the sun god Apollo and his representative on earth, the 'sun king', Louis XIV. Throughout the garden, in literally thousands of statues and ornaments, the solar theme was celebrated.[16] But its primary symbolism resided in the brute domination of the landscape for miles around. Not only did the king's autocratic rule extend over the human population, but nature also had to bow to his command (Figure 3.9). What was lost was any space for the spontaneous manifestation of beauty, the unpredictable, the unplanned, the mysteriousness of what 'simply arises'. Gardening, as envisaged by Le Nôtre, was the expression of the Herculean-Apollonian principles of brute force and intellectual brilliance that have nothing but disdain for the archetypal feminine qualities of mystery, secrecy, and that which lies beyond rational control.

Significantly, the creation of the garden at Versailles was on a scale so grandiose that it required the involvement of the army. It was, in fact, a military operation, whose antagonist was nature. In the early stages soldiers of the Swiss Guard were deployed to dig out the marsh in order to create the Grand Canal, the centrepiece of the garden, and many of them lost their lives due to poisoning from the marsh gas.[17] In the years between 1685 and 1688, when it had been decided that it was necessary to divert water from the River Eure (over sixty miles away) in order to keep the multitude of fountains at Versailles fully operational, up to thirty-seven infantry battalions and six squadrons of dragoons were deployed along with some eight thousand civilian workers in the construction of an aqueduct.[18] Through injuries and sickness (malaria being the main weapon that nature used in her self-defence), the number of casualties was on a suitably military scale. In 1685 (the first year), all two thousand beds of the military hospital serving the project were full, prompting one contemporary commentator to remark that the project would be 'the ruin of the infantry'.[19] The three-year campaign to build the aqueduct ended in a humiliating defeat for the king, since the project had to be abandoned, at a cost of an estimated ten thousand lives.[20]

Versailles has often been linked to the spirit of Cartesianism, which rose to prominence during the seventeenth century. In the writings

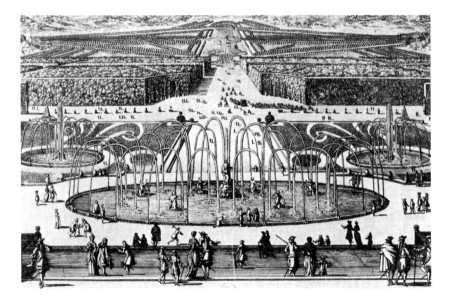

Figure 3.9. The gardens at Versailles, looking west along the central axis, past the pool of Latona, mother of Apollo, towards the canal.

of Descartes, not only was the split in the European psyche between intellect and nature laid bare, but also with it came the need to assert the supremacy of the rational mind over nature, taming her wildness by conforming her to mathematical principles. Descartes' scientific method was founded on the view that through the alignment of human thinking with mathematics, it would be possible for human beings to obtain the kind of exact knowledge which would render them truly 'masters and possessors of nature'.[21] In other words, his scientific thinking was intrinsically technological, being concerned primarily with attaining the means to bend nature to the human will. For this reason it was readily applicable in the aesthetic sphere of garden design, in which the assertion of human ownership was expressed by breaking nature down and reassembling her in accordance with the principles of mathematical reasoning.

Just as the spirit of Cartesianism spread all over Europe, so too did the style of Le Nôtre in gardens far and wide. Unlike the use of the geometrical style in the 'mystical' gardens of Heidelberg and Wilton, the symbolic component was now reduced solely to the assertion of human supremacy over nature. Not only were the later seventeenth and

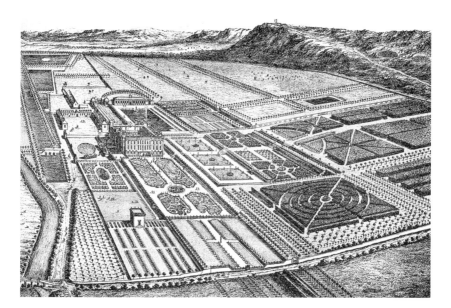

Figure 3.10. The gardens of Chatsworth, Derbyshire in 1707.

early eighteenth century gardens of huge extent, but beyond the garden itself it was considered important that the dominion of the human over the natural world be declared in the imposition of long, straight avenues of trees stretching across the countryside, and the transformation of the landscape into parkland. The garden at Chatsworth, Derbyshire, is typical in this respect (Figure 3.10). One feels, in contemplating such garden designs, an unmistakable coldness pervading the whole. There is something utterly bleak about it. In this great triumph of human reason over disorderly nature, all exuberance, joy and spontaneity seem to have been erased.

The landscape garden

During the seventeenth century, in the midst of the movement to tame nature and bring her under human control, a quiet and largely

Figure 3.11. Landscape by Claude Lorrain (1675).

unnoticed counterpoise emerged. It emerged amongst artists who, at odds with the dominant mood of the time, rediscovered the innate beauty of the landscape. In the paintings of Gaspard Poussin (the adoptive son of Nicolas Poussin), Claude Lorrain, Salvator Rosa and others, the intrinsic beauty of nature was rediscovered. The pastoral scenes that these artists depicted would quite often include a ruined temple amidst a grove of trees, overlooking a lake or river, thus harking back to the lost classical, or pre-classical, age when nature was still the dwelling place of the gods. Here was an appeal to antiquity quite different from that which lay behind the Renaissance recovery of the Roman formal garden. It saw nature as infused with atmosphere, and each painting was, while certainly idealised, nevertheless an evocation of the spirit of place. As such, these artists affirmed precisely what the great gardens of the sixteenth and seventeenth centuries denied (Figure 3.11).

It took some time, but around the beginning of the eighteenth century, a reaction to the formal garden began to take place in England. Joseph Addison wrote in the *Spectator*, on June 25, 1712:

> Our British Gardeners ... instead of honouring Nature, love
> to deviate from it as much as possible ... We see the marks of
> the scissors upon every Plant and Bush ...for my own part,
> I would rather look upon a Tree in all its Luxuriancy and
> Diffusion of Boughs and Branches, than when it is thus cut
> and Trimmed into a Mathematical Figure ...[22]

The reaction was based on the view that nature should not be regarded as artless matter in need of shaping by human art, but that *nature is itself art*, to be appreciated in its own right. As such, it could be a worthy subject for painters to translate onto canvas, and a model for landowners who might wish to follow nature's lead in their gardens. William Kent, who is often credited with being the founder of the English landscape garden, studied and made copies of the work of the landscape artists in Rome, where he lived for ten years. Many English noblemen who visited Italy on the Grand Tour would also have been exposed to the wealth of landscape paintings, especially in Rome. It is significant that in the English landscape movement not only Kent, who inaugurated the English landscape garden at the beginning of the eighteenth century but also Humphrey Repton, who brought it to its final fruition in the late eighteenth and early nineteenth centuries, were both painters. A new bond was being forged between artists and gardeners, and it now entered into the public awareness that gardeners need not so much be *designers* beholden to architectural principles, as *artists* receptive to nature's beauty.[23]

Whereas during the sixteenth and seventeenth centuries gardening was little more than an adjunct to architecture, slavishly bound to architectural design principles, it now began to wrest itself free of both the architectural imperative to impose design on nature and the dictates of geometrical formalism. Gardening now asserted itself as *an art in its own right*. As one contemporary commentator put it, gardening is an art

> ... superior to landscape painting as a reality is to a representation .... The business of the gardener is ... to show all the advantages of the place upon which he is employed; to supply its defects, to correct its faults, and to improve its beauties.[24]

To show all the advantages of the place ... The more that gardening was conceived as an art form independent of architecture, the more the gardener was required to take his cue from *the place* where he worked, rather than just the house or mansion. It was in 1731 that Alexander Pope pronounced his famous injunction to 'consult the genius of the Place'. This became one of the most repeated refrains in texts on the making of landscape gardens throughout the eighteenth century. The gardener, as artist, had to work *with* rather than against nature. And the art of gardening was now conceived as the ability of the gardener to 'improve the beauties' of a place while at the same time staying true to its particular 'genius'.

The garden created in the 1740s by Sir Henry Hoare at Stourhead in Wiltshire was very much influenced by such ideas. It was one of the earliest landscape gardens, and it was, significantly, completely detached from the house (Figure 3.12). Henry Hoare's intention was to create a Poussin-style landscape, now reproduced in three dimensions.[25]

Figure 3.12. The Romantic reaction. Stourhead, in Wiltshire.

It was deliberately set out so that the visitor would experience a series of beautifully composed views. To what extent the spirit of place really was consulted in this and in subsequent landscape gardens is, however, debatable. At Stourhead, as Henry Hoare himself admitted, nature was being recreated in accordance with the idealised 'Arcadian' image of which the landscape painters were so enamoured. But here at least there was a new type of garden, which was based on the idea that the beauty of a place could be enhanced or improved by human intervention, rather than simply being treated as soulless raw material and having an alien design stamped upon it.

At Stourhead, and subsequently at other great houses, certain elements were regarded as essential for the completed picture: a lake (invariably artificial), with its shores planted with beech and cypress trees, various architectural foci, such as a bridge, a pantheon, a grotto and one or more sham ruin, stands of stately trees and gently undulating hills (often created as a result of considerable earth-moving efforts). While in this style of gardening one feels that nature was once again enabled to breathe, it would be more true to say that it was not so much nature that led the way, but a particular image of nature that appealed to eighteenth century sensibilities.

The landscape movement began in the early years of the eighteenth century with just a few disparate voices raised against the excessive formality of the geometrical style, but it achieved a momentum in the 1740s in the work of William Kent and others that soon became a great, unstoppable wave. In the energetic figure of Lancelot 'Capability' Brown, this wave swept the country, and during the years between 1751 and 1783 (the year of his death), one country house after another abandoned its old formal gardens, and replaced them with a new landscaped park. This would typically consist of gently rolling grassy hills (often made by moving tons of earth), planted with carefully positioned stands of trees (elm, oak, beech, ash and lime being his favourites), surrounding a large tranquil lake (made by damming a stream or flooding a marsh). In Figure 3.13, we see how Brown succeeded in transforming Chatsworth's imperious formal gardens into softly contoured parkland stocked with deer.

Unlike his predecessors, Brown had no strong predilection for classical ruins, and his gardens were influenced less by the seventeenth century landscape painters' visions of Arcadia than by his own. In the Brown landscaped park there is no reference to a bygone classical age, but

Figure 3.13. Chatsworth, transformed by Capability Brown.

rather a direct appeal to a new sense of what constitutes natural beauty. The only problem was that, while Brown set out to develop the intrinsic 'capabilities' of whatever landscape he was commissioned to improve, the improvements he made had a tendency to result in an instantly recognisable 'Capability Brown' landscaped park. Heveningham Hall in Suffolk, one of his late commissions undertaken in the early 1780s, has all the familiar hallmarks of an artificially created irregular lake, with the usual stands of trees and placid curvature of hill (Figure 3.14).

By the end of his life, Brown had landscaped in the region of one hundred and forty country estates.[26] Many more estates were landscaped by landowners following his general style, but without the benefit of his expertise. By 1800, across the length and breadth of England, almost every landowner, both large and small, had converted his estate into a landscaped park, following the basic formula employed by Brown.[27]

In the making of the English landscape garden, which was so passionately informed by the desire to sweep away the artificiality and arrogance of the old formal style, a question does unavoidably arise. On the one hand, the landscape movement was a liberation movement

Figure 3.14. Heveningham Hall, landscaped by Brown in 1781.

that freed nature from more than a century of being manacled and maltreated. On the other hand, the new 'nature-friendly' landscapes were normally achieved only through radical interventions involving considerable human labour and expense. Streams or rivers had to be dammed, artificial lakes had to be created, the terrain had to be reshaped and a very large number of trees had to be planted in order to bring about the desired effect. It has been suggested that Le Nôtre and Capability Brown in fact had much in common, for they both worked with a standard formula which they applied to all manner of sites, and they both engaged in massive earthworks in order to imprint their formula upon a given location.[28] In the following centuries, therefore, the question that came to live with increasing intensity in the consciousness of gardeners was: how can we garden in such a way that we can both exercise our human creativity in order to achieve the effects that we want to achieve, while at the same time respecting nature and keeping faith with her intrinsic beauty?

Chapter 4: The Gardener as Artist

The nineteenth century

We have seen that while the eighteenth century landscape movement proceeded from the detached, onlooker consciousness that developed in the Renaissance, it recoiled from the view of nature as little more than raw material with which to construct human designs. Instead, it strove to arrive at a sense of nature's intrinsic beauty, and to enhance it through human intervention. If the older style of garden design rested on an attitude that deemed nature to be bereft of intrinsic meaning and value, the landscape movement arose from a very different attitude. It perceived meanings and values residing within nature as qualities to be grasped and appreciated by the human observer. That this attitude should have guided the sensibilities of the landscape movement, no matter how imperfectly it was actualised, is of utmost significance not only for the history of gardening but also for the historical development of human consciousness. For it marked an important turning point in our whole relationship to nature.

To say that in the eighteenth century nature began to be rehabilitated does not quite express what was occurring; for, along with a new respect for nature, there was also a new sense of human intimacy with her. In their intuitions as to how to improve the sites they worked on, landscapers like Capability Brown, and Humphry Repton after him, undoubtedly felt that – in sweeping away the old formal gardens – they were returning these sites to greater harmony with nature. Their motivation was to work with an idealised natural beauty, to introduce it where it was absent, or to bring it to more perfect expression where it was but imperfectly present. In other words, they felt that they were working *with* nature, towards her improvement. In the latter part of the century, the poets of the Romantic movement articulated what was struggling to the surface of human consciousness: it was not just that they appreciated, indeed reverenced, nature in her own right, but they also grasped that there was the possibility of a new *reciprocal* relationship between human beings and nature.[1] As Coleridge put it, there is 'a bond between nature in the higher sense and the soul of man'.[2] This is something to which we shall return later.

Not everyone, however, felt the way the landscapists and Coleridge felt. While it is true that throughout the nineteenth century the landscaped park, with its implicit deference to nature's inherent beauty, continued to be the setting for the large country house, the older impulse that saw gardening in terms of human beings exercising their creative mastery over nature was far from dead. The formal garden soon began to creep back again in the areas immediately surrounding the country house. The deeply ingrained habit of thought that saw nature as raw material to be shaped and designed to fit our human ideas of what is beautiful, was completely in tune with the spirit of industrial England. Both science and industry rested on this view of nature, and it not surprisingly asserted itself in nineteenth century garden aesthetics. Furthermore, the splendour of the British Empire, with all its backward

Figure 4.1. The Italianate Garden at Eaton Hall, Cheshire in 1857.

glances to Imperial Rome, required majestic gardens to complement it. The landscaped park did not sufficiently represent the imperial spirit, whereas the architectural style of the Renaissance garden did. And so, in Victorian England, the 'Italianate' style of garden became the new standard in public spaces and around the homes of the wealthy.

An example of the Italianate style can be seen in Figure 4.1. It once again introduced architectural features such as terraces, stairways and balustrades into the garden. Pagodas, ornamental pools and fountains, urns and statuary all returned. So too did the swirling patterns of the parterre, along with the use of different coloured gravels to add to the overall effect. Topiary also once more came back into vogue. From the 1840s onwards, the use of massed flowers in parterres to create elaborate coloured patterns in low beds, became immensely popular in large gardens and municipal parks. The abolition of the tax on glass in 1845 meant that a much wider range of plants could be grown in glasshouses. While this generated greater awareness of and interest in flowers, which had been totally neglected by the Landscape Movement, it also allowed the notion to be fostered that plants could be grown simply to serve as pigments in floral pictures. Thus arose the idea of 'carpet bedding' (strictly speaking the use of dwarf foliage plants to create designs) and subsequently 'picture bedding', (which used both foliage and flowering plants to create designs). Favoured especially in municipal parks, the classic design was the municipal coat of arms, surrounded by a suitable motto. The genre has continued to this day, one of the most famous and persistent designs being that of the Edinburgh floral clock, which first appeared in 1903, and has continued to be recreated each year to the present time (Figure 4.2).

The floral clock, along with countless other carpet or picture bedding designs, needed painstaking efforts in order to achieve the required result, from the preparation of the soil to the drawing out of the plan on the ground, and then filling it in with hundreds if not thousands of plants. Like the parterre, like topiary too, carpet and picture bedding have to be acknowledged as a valid art form within their own prescribed limits, and one cannot deny the care and love that were – and still are – bestowed upon the production of such pictures. We must nevertheless question the relationship to nature that underpins this style of gardening, for it rests upon the reduction of the plant to nothing more than a colour element in a greater picture, very literally conceived as an object such as a coat of arms, a clock, a steam train or whatever. The literalism of such floral pictures is what makes us feel that they mainly serve to bolster a sense of human assurance in the midst of the city, through the enslavement of the plant kingdom to not very elevated human aesthetic tastes. These practices arise out of, and indeed accentuate, what is at root a profound sense of alienation from the natural world. Nature is regarded solely as *Natura naturata*, lacking any inner life, lacking any quality of soul. Every plant is just a thing, an object amongst other objects, whose characteristics are valued solely for their usefulness to the floral planting scheme.

And yet, how many of us practising gardeners do not fall into precisely this attitude? We may turn our noses up at municipal floral pictures, but is there not a tendency in all of us to view plants as simply colour elements in our garden designs? What is taken to an extreme in one garden style is what we all find ourselves doing to some degree, because we are all tainted by the same tendency to relate to nature as an object 'out there,' rather than to reach into her inwardness. How many of us have even begun to consider what a garden might be, were it an

Figure 4.2. The Edinburgh Floral Clock, 1910.

expression of our relationship to that interior dimension of the natural world that the medievals referred to as *Natura naturans*?

We shall return to these thorny questions shortly, but first let us continue to follow up the nineteenth century desire to find a compromise between the informality of the landscaped park and the formality of the Italianate style. This was reached in the so-called 'gardenesque' style, which was particularly well suited to the large suburban garden. The 'gardenesque', a term coined by John Loudon in the early part of the nineteenth century, was essentially formal; but within the overall context of order and control it allowed irregular planting and a degree of informality as well (Figure 4.3).

One of the chief aims of the gardenesque was the display of specimen shrubs and flowering plants. During the nineteenth century, more and more plants from far off continents were introduced into Europe, and enthusiastically embraced in the gardens of England. Garden magazines began to be published (Loudon's *The Gardener's Magazine* started up in 1826, and was soon followed by others), and books on gardening proliferated, appealing to the burgeoning middle classes who had sufficient land and leisure to establish an ornamental garden. The introduction of the mechanical lawn mower in the 1830s made it much easier for that key element of the landscaped garden – large areas of green turf – to become

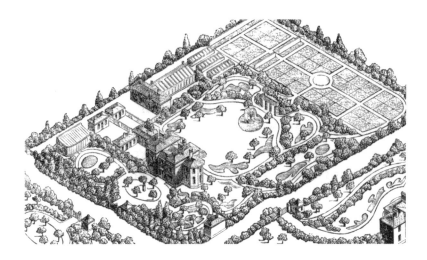

Figure 4.3. John Loudon's design for a suburban villa.

the backdrop of the suburban garden, though now on a much smaller scale. As we have seen, in 1845 the tax on glass was abolished, which meant that glasshouses became more affordable, allowing far more people to grow a wider range of plants. Thus the creative impulse in gardening began to pass in the nineteenth century out of the hands of the wealthy landed gentry and their successors, the industrial magnates, into the hands of the middle classes. The small garden, rather than the country estate, increasingly became the locus of creative gardening.

Loudon defined the gardenesque as 'the production of that kind of scenery which is best calculated to display the individual beauty of trees, shrubs and plants in a state of nature'.[3] In such a conception, human control is paramount, but it is exercised in the service of exhibiting nature's beauty. Thus the gardenesque ideal precariously held the tension between two contrary impulses: on the one side to control and direct nature and make her conform to human aesthetic requirements, and on the other side to work with nature in such a way as to allow her intrinsic beauty to be displayed. Underlying this tension there lived not only the question of how to achieve a greater reciprocity with nature, but also a further, deeper question about the purpose of the garden that no one had yet been able satisfactorily to formulate, let alone answer. Towards the end of the nineteenth century, this question lived in people more and more strongly, and it may be expressed as follows: Does the garden exist for the sake of nature or does nature exist for the sake of the garden? In the late nineteenth century this question was finally answered in different ways by two brilliant and hugely influential figures: Gertrude Jekyll and William Robinson.

Gertrude Jekyll

Gertrude Jekyll began her adult life with the hope of becoming a painter, studying at the Kensington School of Art in the early 1860s. Coming from a cultured middle class background – her father was a retired army officer, her mother a musician – she moved with ease in literary and artistic circles. In time, however, she was forced to abandon painting for her other passion – gardening – because of her weak eyesight. In the course of her long life she had an enormous influence on the way in which gardening was conceived and practised. Both through her garden designs and through her prolific writings, she pioneered a style of gardening that lives on into the present day.

Jekyll's approach to gardening was from the beginning that of a painter. But she was about as far from the landscape painting tradition as one could imagine. Jekyll's aim was less to reproduce some idealised view of nature than to use the colours, shapes and textures of the plants themselves to create a beautiful composition. Her interest was thus less in landscapes than in what could be achieved in the smaller garden, in which the characteristic forms of individual plants would, through judicious planting, produce beautiful and harmonious effects. The garden picture would for her be something far more refined than the literalistic floral picture, and again more akin to the contemporary Impressionist style than to the landscape artists of the seventeenth century. She wrote:

> In setting a garden we are painting a picture – a picture of hundreds of feet or yards instead of so many inches, painted with living flowers and seen by open daylight – so that to paint it rightly is a debt we owe to the beauty of the flowers and to the light of the sun; that the colours should be placed with careful forethought and deliberation, as a painter employs them on his picture, and not dropped down in lifeless dabs.[4]

For this kind of gardening to work, the gardener has to have a great deal of knowledge about the different habits of plants, their soil requirements and manner of growth. Everything has to be carefully thought out, with each plant chosen and then rightly positioned for the part it is to play in due season. Nothing is left to chance, and every effect produced is produced deliberately.

In Figure 4.4, the red section of the herbaceous border at Munstead Wood, where Jekyll lived, is shown in all its glory. For Jekyll, the ideal progression of colour along the herbaceous border was from the cool colours at either end (blues and greys, pale yellows and pale pinks), through purples and violets to the climax of oranges and reds in the centre.[5] To gaze at such a 'garden picture' (a favourite phrase of Jekyll's) is to experience oneself in a different world from the world of formal parterres and municipal 'floral pictures'.

The hallmark of a Jekyllian garden was that it would be eminently transferable to canvas. Just because the gardener was an artist painting 'living pictures with living flowers', painters would feel drawn to portraying what they saw in the garden.[6] Figure 4.5 is one such painting of the blue-grey flower border at Munstead in October. In this, one sees a reversal of the relationship between painters and gardeners that characterised the eighteenth century landscape movement. For the first time, the gardener now confidently asserted her entitlement to be regarded primarily as an artist, whose work would then become a subject of interest to the painter.

Jekyll had an exceptional sensitivity to nature, and loved the humblest wild flowers as much as the showier garden plants. In the garden, however, individual plants would have to serve as elements in a composition that it was the gardener's role to devise. This composition,

Figure 4.4. Red section of Gertrude Jekyll's flower border, c.1912.

of course, was no literalistic picture as in the municipal planting tradition, but rather took its lead from the characteristics of the plants themselves. Through the gardener's knowledge of and sensitivity to the plant kingdom, the planting was orchestrated so as to bring about an outcome to which each plant contributed in its unique way. It was in this sense of taking creative responsibility for making a 'garden picture' that Jekyll saw the gardener as an artist. In the garden, the gardener's palette was made up of all the different plants that she or he 'painted' with. Each plant was therefore valued primarily for the contribution it made to producing the desired overall aesthetic effect.

In the white garden at Hidcote (Figure 4.6), we see another example of Jekyllian gardening principles in action. Jekyll herself did not design this garden but it conforms closely to her ideas. In the white garden, nature is civilised, and is brought into a florescence of elegance and grace through the gardener's sense of the beautiful. In this and other similar planting schemes, the garden is made to fit around the house, and nature is thus accommodated within the human world.[7] In a

Figure 4.5. The blue-grey flower border at Munstead. Painting by Helen Allingham.

sense Jekyll *humanises* nature. Everything is carefully planned so as to assimilate nature to human culture, so that one scarcely has the experience of stepping outside the human domain at all. For Jekyll the house and garden form a unity, and it is significant that many of the designs that Jekyll made were in collaboration with the architect Edwin Lutyens. Jekyll did not believe, however, that house and garden should be artificially isolated from the larger natural landscape, and in the case of country gardens she recommended that, as the garden approached the countryside, indigenous trees and under-planting should be carefully selected.[8] Thus by subtle human intervention, the garden would merge seamlessly into the surrounding landscape.

If Jekyll strove to lift nature to a higher level of refinement by incorporating the natural within the human cultural and aesthetic realm, then she also understood that this would not happen unless human beings re-educated their senses and developed a keen appreciation of the innate beauty of the natural world. It is the fact that Jekyll encourages us to look at plants in a completely fresh way that gives to her writings a lasting value and makes her into a pioneer working towards the future. Consider, for example, the following description of the humble wood sorrel:

Figure 4.6. The white garden at Hidcote.

The white flower in the mass has a slight lilac tinge; when I
look close I see that this comes from a fine veining of reddish-
purple colour on the white ground. White seems a vaguely-
indefinite word when applied to the colouring of flowers;
in the case of this tender little blossom the white is not very
white, but about as white as the lightest part of a pearl. The
downy stalk is flesh coloured and half-transparent, and the
delicately-formed calyx is painted with faint tints of dull
green edged with transparent greenish buff, and is based and
tipped with a reddish-purple that recalls the veining of the
petals. Each of these has a touch of clear yellow on its inner
base that sets off the bunch of tiny whitish stamens.[9]

In this way Jekyll urges us to *really look* at plants. Gardening becomes
a kind of training in how to make ourselves into ever more sensitive
instruments, capable of noticing and appreciating the subtlest qualities
in plants.[10] This faculty in her was not restricted simply to the sense
of sight, but extended also to the sense of smell, for she could name
different varieties of rose with her eyes shut from their scent alone. It
extended as well to the sense of hearing. She was, for example, able to tell
what trees she was nearby simply through the sound of the wind in their
leaves, and wrote of the different 'voices' of the birch, oak and chestnut.[11]

Nevertheless, while Jekyll sets us on the path to the re-sensitisation of
our whole relationship to the natural world, her avowed purpose was not
in the end to evoke the inner mysteries of nature. She always laid stress
rather on 'gardening for beautiful effect'.[12] Her avowed aim was primarily
to provide a setting in which human beings could relax and at the
same time be uplifted by their experience of being in a (literally) highly
cultivated environment. The garden, in Jekyll's view, exists for *human
beings*, and all the skill of the gardener is to utilise nature to promote
human wellbeing. In 1899, she summed up her philosophy as follows:

I hold firm the belief that the purpose of the garden is to
give happiness and repose of mind, firstly and above all other
considerations, and to give it through the representation of
the best kind of pictorial beauty of foliage and flower that can
be combined or invented.[13]

William Robinson

The other hugely influential figure for the development of the modern
garden was William Robinson, a contemporary and friend of Gertrude
Jekyll. Unlike Jekyll, Robinson was of humble origins. Born in Ireland,
he worked first of all as a garden boy on a large country estate, then as
an apprentice, and became foreman in a smaller rural garden in his early
twenties. At the age of twenty-three he left Ireland for England, to work
in the gardens of the Royal Botanic Society at Regents Park, where
his talents were quickly recognised. During his lifetime, Robinson
travelled widely, both in the British Isles and abroad, in France and
America. Like Gertrude Jekyll, he was also a prolific writer. Unlike
her, he was a fiery character, self-educated and with very strongly held
opinions, which he expressed with great passion.

Robinson shared with Gertrude Jekyll the view that gardening
is an art, and he too thought of gardening as 'creating pictures'.
But his approach to creating garden pictures was quite different
from Jekyll's approach. Whereas Jekyll sought to raise nature to a
higher level of beauty by means of the gardener's artistic skill and
sensitivity, for Robinson it was Nature (always spelled with a capital
'N') who was the supreme artist, to whom the gardener should be
a helper or assistant. He saw the gardener's task as being to enable
Nature to display herself as unselfconsciously as possible, and to
do this it was necessary to keep faith with and stay true to Nature,
and to 'trust her guidance'.[14] The gardener, in the role of assistant,
should try to bring to expression all that is best and most beautiful
in Nature. He said:

> The aim should be never to rest till the garden is a reflex of
> Nature in her fairest moods.[15]

For Robinson, Nature was more than just the sum total of plants,
soils, rocks and animals. Nature was an entity in her own right, a
veritable being, the 'Universal Mother' of whom Homer sang, and
with whom every gardener is in direct relationship.[16] In this respect,
we are reminded of the medieval recognition of the goddess Natura,
and of the idea that Nature has an inward as well as an outward

mode of being. Robinson was a practical man, not a philosopher or a mystic, but he nevertheless had an innate and unschooled sense of the mystery of life, and a reverence for Nature that informed his whole approach to gardening. While his conception of the garden does not have the same mystic resonances of the medieval *hortus conclusus*, in which plants are valued for their religious symbolism and for the moral lessons they may teach us, yet the direction of his thinking about gardens was to see them as the means by which Nature could be felt as a sublime presence.

Robinson was a pioneer of the herbaceous border, and has rightly been called the 'father of the English flower garden'. But where his approach to gardening differs from that of Jekyll is perhaps best seen in his advocacy of the wild garden. In his idea of the wild garden, Robinson more closely defined how he saw the relationship between the human and the natural spheres. The wild garden was by no means simply an area left completely untended. Rather, it required the most delicate and deferential human intervention, in harmony with the spirit of a place, and with the aim of enhancing its intrinsic beauty. This was to be achieved by naturalising plants, introduced by the gardener from elsewhere, but with all due sensitivity to the inherent meaning of the site where the garden was being created. Referring to the gardener as an artist, he wrote:

> In such a garden it might be clear that the artist had caught
> the true meaning of Nature in her grouping, without
> sacrificing anything of value in the garden.[17]

An idea of what he was aiming at can be seen in Figure 4.7, where the gardener's input is more one of benevolent management and guidance than the imposition of an overly civilising human aesthetic.

Robinson was the first modern gardener to express the idea that the art of gardening has a *sacred*, not just an aesthetic, dimension to it. This conception springs quite naturally from his feeling that Nature is a living, and indeed a divine, being. Implicit in his approach to gardening is the assumption that in each and every garden, the whole of Nature is involved. Thus one does not simply garden with the particular elements – soil, shrubs, herbaceous plants, trees, and so on – that go to compose

Figure 4.7. A Robinsonian wild garden in spring

an individual garden. No matter how varied or unvaried these might be, they do not constitute the primary relationship that the gardener has to foster, but rather they are instrumental to it. The primary relationship is to Nature, whose beauty can become more or less manifest depending on how sensitively we garden. He writes:

> A garden should be a living thing: its life not only fair in form
> and lovely in colour, but in its breath and essence coming
> from the Divine.[18]

In this statement, we meet a very clear, albeit tentative, articulation of a way of gardening that, while reminding us of the past, really belongs to the future. Gardening has the capacity to be more that just art, as Jekyll conceived it, fair in form and lovely in colour; it also has the capacity to be *art that mediates the divine.*

Figure 4.8 shows the wild garden at Heaselands, in Sussex. Here we can perhaps glimpse the kind of thing Robinson was aiming for. What we see (and would smell were we there) is essentially a 'breath and essence coming from the Divine', yet all this beauty has been achieved not by unaided Nature, but also through creative human involvement. It is only through the co-operative effort of human beings with Nature that this kind of Paradisal environment can come into being. We notice, too, that this wild garden is outside the borders of a garden as conventionally understood. Robinson believed there was no need to restrict gardening to conventionally demarcated garden areas, because for him the whole world was a garden.[19] Thus the beauty of fields, meadows and woodlands could all be enhanced through such sensitive human participation.

Figure 4.8. The wild garden at Heaselands, Sussex.

Robinson never set out a systematic philosophy of gardening, and the vast majority of his writings, although frequently laced with his vigorously held opinions, are emphatically pragmatic. Furthermore, his own gardening did not always entirely bear out his principles. The garden at Gravetye Manor surprised many of its visitors with its formal beds around the house. His vital contribution, however, lay in his understanding that what was of crucial importance was the attitude of the gardener to Nature. The purpose of gardening was, for him, to enhance Nature's glory rather than to display human aesthetics. And in this attitude lies the seed idea of gardening as a sacred art.

Chapter 5: Gardening as a Sacred Art

The guiding light of Monet

The artist, Claude Monet, was a contemporary of Gertrude Jekyll and William Robinson. All his life he was as much a gardener as he was a painter, subscribing to horticultural magazines and encyclopaedias, and ordering seeds from around the world.[1] A large number of his paintings were of the gardens that he created in the rented properties where he lived. Once ensconced at Giverny (where he moved in 1883), the garden was the main subject of his work as a painter, as well as engaging much of his creative energy as a gardener. Monet once said to a journalist: 'I perhaps owe it to flowers for having become a painter.'[2] This statement shows how Monet's life path wonderfully complements that followed by Gertrude Jekyll, who might have said: 'I owe it to painting for having become a gardener.' Both Jekyll and Monet straddle the world of gardening and painting. Living at the threshold of the twentieth century, they both exemplify the modern conception of the gardener as artist, or the artist as gardener.

At Giverny, Monet's *coup de grâce* was the creation of the lily pond, surrounded by flowers, trees and flowering shrubs in the lower part of the estate (Figure 5.1). When I first visited it in the 1980s, I could not help but be struck by the powerful atmosphere that seemed to pervade the garden. The other visitors – and there were many, for it was summer – were either completely silent or else speaking to each other in low whispers. It was as if we were all in a cathedral, only the cathedral was outdoors, and instead of being made of glass and stone it was made of water and flowers.

My experience at Giverny was not simply that this garden was a beautiful 'picture', like a Jekyllian garden, to be observed and appreciated in the manner of a detached onlooker. It is, of course, a

stunningly beautiful garden, but there was something else that also
entered into the experience. All of its profusion of beauty seemed to
be held in a magical atmosphere that pervaded the whole garden. And
it was this atmosphere, rather than any particular grouping of plants,

Figure 5.1. The lily pond, Giverny.

which was so arresting; for through its atmosphere, one was aware of the garden less as an object than as a *living being* which surrounded one on all sides. And to the extent that one felt the 'beingness' of the garden, all the outwardly visible phenomena (the sunlight on the water, the round-leaved lilies, the dragonflies, the proliferation of flowers and foliage on the waterside) seemed to be pointing towards another level of life – the mysterious and invisible inner life of nature, so potently present there.

Monet himself spoke and wrote very little about what he thought he was doing as a painter, even less as a gardener. But he did have some very articulate contemporary admirers, who were able to put into words what they felt Monet was trying to do in his artistic work. One of these was Octave Mirbeau. Mirbeau was one of Monet's most committed supporters from the 1880s onwards, and saw him as a secular mystic who captured 'the dream which is mysteriously enclosed in nature'.[3] He felt that what Monet was doing was accessing nature's thoughts, moods and reveries, and so was able to transcend the mere 'naturalism' of the objectivising consciousness, while at the same time remaining true to the inner life of nature. Mirbeau wrote:

> He [Monet] reveals the impalpable, the ungraspable in nature ... its soul, the thoughts of its mind and the beating of its heart.[4]

While Mirbeau was referring to Monet's painting, exactly the same thing could have been said about his gardening. What Monet was trying to do was to sink himself ever more deeply into the phenomenal world, to penetrate nature's creativity through his own creative activity. It was with his artist's eyes, his paintbrush, and also with his garden-making, that he strove to arrive at the very source of beauty. He once wrote in a letter to a friend that what he sought as an artist, above all, was 'the *enveloppe*, the same light spreading everywhere'.[5] The word *enveloppe*, as Monet uses it here, approximates to the English word 'atmosphere', for it is what surrounds or wraps itself around the individual phenomena, bathing them in 'light'. Like the atmosphere of a garden, the *enveloppe* cannot itself be directly represented because it is *the condition of things becoming manifest*. It is the unrepresentable matrix within which, and through which, what we perceive becomes visible.

For Monet the garden was a place of deep contemplative engagement with nature. As a gardener and as a painter, he was dedicated to refining his own perceptual experience to the point at which he was able to break through to awareness of this other level of nature. This level is reached only when one goes beyond the objectivising perception of things 'outside' oneself, and experiences the *originating source* of the perceptual act, which is neither wholly on the side of the observer nor wholly on the side of the observed. We can sense Monet's utter dedication to this in his many 'series' paintings, in which he returns again and again to the same subject, placing himself with his canvas in exactly the same position, in order to capture it under different light conditions. The famous paintings of the Japanese bridge at the lower garden in Giverny are one example of this commitment to the intensification of the act of perception (Figure 5.2). He painted this bridge at least fifteen times.

The garden that Monet created at Giverny was also his studio. It was made with the very specific intention of evoking precisely what he wanted to capture in paint. This means that the purpose of the garden was to conjure forth the evanescent *enveloppe*, that translucent 'atmosphere', which still stops people in their tracks when they visit Giverny. Because of its effect on visitors, the garden has become a kind of sacred precinct. It is not that it is a shrine to Monet, but rather – through its powerful 'spirit of place' – it has become a shrine to the numinous source of nature's exuberant creativity. At Giverny one feels that through the intensive application of human imagination and labour something more than human has been enabled to show itself. In this respect, the artist-gardener has created not so much a 'garden picture' in the Jekyllian sense, as what could best be described as *an icon*.

The garden as icon

Icons have been defined in many different ways, but broadly speaking one might say that icons are images that have the capacity to make manifest that which is essentially invisible and non-manifest. Michel Quenot explains that the icon is:

> a visible sign of the invisible presence; therefore it eclipses both the painter and the spectator by the very fact of its transcendent element, the sacramental presence.[6]

In the icon, then, what is represented becomes alive in the representation, so that the representation is no longer simply an object to be gazed upon, but is also experienced as a living presence. So while it is composed of

Figure 5.2. One of Monet's studies of the Japanese bridge, 1899.

recognisable elements of our visible world, it employs them in such a way that they become mediators of an invisible reality. Apprehended in this way, as a presence rather than merely as an object, we relate to it differently. It then becomes accessible to us: not only can we communicate with it but it can communicate with us.

If the icon is not simply an object, it is because it exists at the threshold between worlds, where the visible touches the invisible, and where the created meets the creative. Icons have been termed 'windows into heaven' for this reason. The heaven of the garden, however, is not somewhere else. As we have seen in the old teaching that nature has two aspects, *Natura naturans* and *Natura naturata*, the term *Natura naturans* denotes the creative and formative forces that stand behind the visible world of sense-perceptible forms. By contrast, the term *Natura naturata* denotes all the visible effects of nature's creative activity, namely the manifest world of mineral, plant and animal forms that we see around us. *Natura naturans* was implicitly understood to be the world of spirit, no different from God, and it was explicitly identified as such when the term was first introduced in the early thirteenth century.[7]

The distinction lived on through the Renaissance and even survived into the seventeenth century where it played a prominent part in the writings of Spinoza.[8] When we come to the nineteenth century, we find it utilised by German idealist philosophers, and their British exponents, like Samuel Taylor Coleridge. Coleridge clearly grasped the import of the distinction for the artist, dismissing the mere copying of nature (i.e. *Natura naturata*) as 'idle rivalry' and 'emptiness'. According to Coleridge, the true artist needs 'to master the essence, the *Natura naturans*, which presupposes a bond between nature in the higher sense and the soul of man.'[9] What Coleridge saw was that the concept of *Natura naturans* opens up the possibility of a new creative engagement with nature, based on the experience of a communion between the inner life of human beings and the inwardness of nature.[10]

The implication of Coleridge's statement is that, as artists, gardeners should not simply concern themselves with the forms, textures and colours of plants and the aesthetic permutations of planting schemes. While knowledge and skill in these areas is of course important, there is more to being a gardener than just acquiring practical knowledge and a degree of aesthetic refinement. There is something else that the gardener also needs to develop. And this is the ability to position oneself at the very margins of the visible world, where the productive forces within nature unfold themselves into outwardly manifest forms. This 'location' – as Monet so well understood – is both interior to, and yet is also to be understood as *surrounding* the phenomena, for in reality they are *in it* rather than it being within them. It is an interior location in the sense of holding the inner identity of a site in which the different elements of the garden are immersed, and which gives to each garden its particular character and atmosphere. One could say that here, in the embrace of *Natura naturans*, is the garden's 'soul', and it is the task of the gardener to form a living bond with it, so that it informs all the decisions that he or she takes.

We all know of gardens that are spectacularly well planted and well tended, providing an impressive display through the seasons, but nevertheless feel soulless. Somehow, the disparate elements of the garden do not form a coherent whole. There is no atmosphere. This is because all the focus of the gardener is on *Natura naturata*, and the intangible quality of 'atmosphere' that a gardener needs also to cultivate, has been ignored. The key to gardening 'atmospherically' is to abandon the pretension that one is alone the creative architect and designer of the garden. One endeavours instead to be a collaborator in a greater creative process. Rather than the gardener feeling that he or she is the creator of the garden, exclusively responsible for bringing together all its different elements, it is possible to reach the experience that *the garden is composing itself* through the gardener's awareness of what needs to be done in the garden. This presupposes an attunement of the gardener's imagination to the spirit – and spirits – of the place, which seek through the gardener to become ever more fully alive and 'present' in the garden. And so the garden, as a work of art, results from a dialogue between the gardener and the spirit (perhaps one could equally well say 'soul') of the garden that is seeking to attain a fuller expression.

The poet, William Blake, well understood the shift in attitude that this entails. In his prophetic poem, *Milton*, he describes how the muse of pure poetry, which appears to his visionary imagination as a young and winged feminine being called Ololon, appears to him in the garden of his cottage at Felpham. Blake addresses this apparition with the following words:

> What is thy message to thy friend?
> What am I now to do?
> ... Behold me
> Ready to obey.[11]

Blake's picture of the encounter is shown in Figure 5.3. What the encounter teaches us is the importance of the willingness of the artist (in this case William Blake) to put his creativity at the service of something beyond the ego, thereby throwing himself out of his own self-centredness. The artist then enters into a productive dialogue with that which hovers beyond the threshold of the visible world, and which is searching for a conduit through which it can come into manifestation. Just as Blake puts himself at the service of his muse, in the same way the gardener can put him- or herself at the service of the spirit of the garden, asking as Blake asked: 'What am I to do now?' or 'What wants to happen here?' This is the opposite of working from a preconceived plan. It is rather an effort to participate in a dynamic and living process, in which the gardener's role is not so much to design and create the garden

Figure 5.3. William Blake encounters the spirit Ololon.

in the image of the *gardener's* creative genius, but more to promote the spirit of place towards a fuller expression *its* genius.[12]

Such an approach to gardening requires that one bring into one's own consciousness what lives inwardly as the creative potential – or the creative desire – of the garden. It presupposes a devotional approach to nature, in which one's whole purpose is to attune one's own imagination and creativity to 'the dream which is mysteriously enclosed in nature' as this lives within the defined space of the garden.[13] Thereby, one steers the garden towards becoming an icon – a living image, as William Robinson put it, 'not only fair in form and lovely in colour, but in its breath and essence coming from the Divine'.[14] In Figure 5.4, something of this may perhaps be felt, although photographs are never able adequately to convey the actual experience of being *in* a garden. In this North Oxford garden, where I worked for many years, one could have the sense of spiritual presence, or presences, as if the nature spirits and the angels were nearby, for the veil between *Natura naturans* and *Natura naturata* seemed at times to be almost transparent here.

At the beginning of this book I suggested that the garden calls to us from the future, where it exists as a still unrealised ideal. It may prove

Figure 5.4. A garden in North Oxford.

ever more difficult to respond to this call especially if, as we move into the future, the keenness of our senses and our sensitivity to interior realms is blunted by the spell of sterile technologies, whose tendency is to draw us further and further away from direct relationship with the living world of nature. But no matter how bedazzled, seduced and distracted we may become, the call of the garden will not readily be silenced, for it beckons us toward the necessary counterbalance and corrective to a consciousness that seems bent on veering away from the real. And so we may feel ourselves inwardly inclined to answer it not just for the sake of nature but also for our own sakes, for we are not complete as human beings unless we are able to stand in the world and experience sacred presence.

The challenge, then, in our secular times, is to find the way to re-sacralise our work as gardeners. Gertrude Jekyll, William Robinson and Claude Monet each in their different ways pointed to new forms of creative engagement with nature that would lead, through devotion to beauty, to a re-sanctification of the world. There can be no going back, no resurrection of the past, but there are viable ways forward. Whereas in antiquity the gods were directly experienced in nature and one ignored them at one's peril, today we are in a very different position: for most people the gods and nature spirits are no longer interwoven with their experience of nature. In order once again to consciously relate to this invisible realm, we have to work at re-sensitising ourselves to it. This can be done both by making a deliberate effort to re-attune ourselves to the spiritual qualities that infuse the sensory world that surrounds us, and at the same time by becoming *creatively engaged* with this more inward and hidden dimension of nature through our gardening. Thus we open the way once more to letting the divine back into our world, and we come to see that gardening has the possibility of opening a window to the spirit. And so our gardens may come to feel more and more like icons, mediating that which is numinous in nature. To the extent that we are able to achieve this, our gardening may at last begin to mature into a sacred art.

Illustration Sources

Chapter 1: The Garden in Antiquity

1.1. Assyrian garden surrounding a hilltop temple. Seventh century BC. Marie Luise Gothein, *A History of Garden Art* (London: J.M. Dent and Sons, 1928), vol.1, Fig. 34, p.37.

1.2. Egyptian temple garden, eighteenth century BC. Tomb of Neferhotep, Thebes. From J. Gardner Wilkinson, *The Ancient Egyptians: Their Life and Customs*, vol.1 (London: John Murray, 1853), p.27.

1.3. The goddess Nut appears in her sacred tree, the *sycomorus ficus*. Vignette to Chapter 59 of *The Book of the Dead*. Papyrus of Ani. From E. A. Wallis Budge, *The Book of the Dead* (London: Kegan Paul, Trench Trubner, 1923), p.204.

1.4. Archetypal Egyptian garden, with an abundance of plants surrounding a sacred lake. Fifteenth century BC. Tomb of Rekhmire, Thebes. Gothein, *A History of Garden Art*, vol.1, Fig. 20, p.20..

1.5. The garden of Nebamun, from his tomb at Dra Abu el-Naga, West Thebes. Eighteenth Dynasty. London, British Museum, no. 37983. Courtesy of the British Museum.

1.6. The human and divine worlds meet in the sacred garden. Papyrus of Nakht. Late fourteenth century BC. London, British Museum papyrus 10471, sheet 21. Courtesy of the British Museum.

1.7. Two Pan figures, each holding goats, while above three nymphs dance. From the Enneakrounos or 'Nine-spouted Fountain House' on the south east corner of the ancient Agora, Athens. *c.*520 BC. From Jane Harrison, *Prolegomena to the Study of Greek Religion* (London: Merlin Press, 1962), p.290, Fig. 73.

1.8. Two satyrs and a maenad dance wildly amongst trees. Athenian red figure vase, fifth century BC. Author's drawing.

1.9. Garden of the Philosophers, thought to represent Plato's Academy. Mosaic from the House of T. Siminius Stephanus, Pompeii. Early first century BC. Photo by Sergey Sosnovskiy.

1.10. Wall painting from the Villa of Livia, Primaporta, near Rome. First century BC. Livia was the wife of Augustus Caesar. Museo Nazionale Romano, Rome. Photo: The Bridgeman Art Library.

1.11. Drawing of a fresco, showing an ornamental pool surrounded by arbours, fenced enclosures and pergolas arranged symmetrically. Auditorium of Maecenas, Rome. Late first century BC. From C.H. Daremberg and Saglio, E.D.M., *Dictionnaire des Antiquités Greques et Romaines* (Paris, 1900).

1.12. Drawing of a fresco, in which all the garden elements are carefully arranged within a symmetrical design. Early Empire period, Pompeii. Gothein, *A History of Garden Art*, vol.1, Fig. 87.

1.13. A grotto, with a *nymphaeum* at its entrance, and a pergola above. Fresco on the wall of the Villa of Publius Fannius Sinistor, Boscoreale. New York, Metropolitan Museum of Art. © Photo: SCALA, Florence.

1.14. The legacy of Rome: topiary in an eighteenth century Viennese garden. Engraving of the garden of Adam von Lichtenstein, 1724.

Chapter 2: The Garden in the Middle Ages

2.1. Typical Islamic garden. Mughal miniature, *c.*1603. Bibliothèque Nationale de France. Author's drawing.

2.2. The Garden of Eden as archetype on which the medieval *hortus conclusus* was based. Dutch engraving, 1503. Sir Frank Crisp, *Mediaeval Gardens* (London: Bodley Head, 1924), vol. 1, Fig. 116.

2.3. The 'Mary Garden' as a *hortus conclusus* on the model of the Paradise garden. *Speculum Humanae Salvationis* (printed edition, *c.*1475), Chapter 3. From Crisp, *op. cit.*, vol. 1, Fig. 11.

2.4. The goddess Natura sustains the earth with her milk. Late fifteenth century illuminated manuscript of Aristotle's *Physics*, 1.1. Austrian National Library, Vienna, ms. phil. gr.2, sheet 1,r. ÖNB/Vienna, E19.259–C(D).

2.5. The degrees of Nature. Carollus Bovillus, *Liber de Intellectu* (Paris, 1510).

2.6. The poet and composer Guillaume de Machaut sits writing beside a spring, within a walled garden, planted with both trees and flowers. Bibliothèque Nationale, Ms fr.1586, f.30v. *c.*1360. Author's drawing.

2.7. The virgin as the enclosed garden. *Speculum Humanae Salvationis*, *c.*1430. Det Kongelige Bibliotek, Denmark, GKS 79 2°, 24, r. Author's drawing.

2.8. Mary as Rose of the World (*Rosa Mundi*). Spanish woodcut, late fifteenth century. Biblioteca Universitaria, Valencia.

2.9. The dominant figure of the Virgin Mary in the *hortus conclusus*. Wood engraving, 1475. From Crisp, *op. cit.*, vol. 2, Fig. LXII.

2.10. 'The Garden of Paradise' by an unknown Master of the Middle Rhine, *c.*1415. Städel Museum, Frankfurt am Main.

2.11. The *hortus conclusus* as a garden of love. Fifteenth century woodcut. From Crisp, *op. cit.*, vol. 2, Fig. CCIV.

Chapter 3: From the Renaissance to the Eighteenth Century

3.1. Training the new perspectival consciousness. Woodcut by Albrecht Dürer, 1538. From Willi Kurth, ed., *The Complete Woodcuts of Albrecht Dürer* (New York: Dover, 1963), Fig. 340.

3.2. North Italian landscape, by Dürer, *c.*1495, entitled *View in the Val d'Arco*. Musée du Louvre, Paris. The Bridgeman Art Library.

3.3. Garden in the early Italian Renaissance style. German engraving, 1641. From Crisp, *op. cit.*, Fig. 159.

3.4. The gardens of the Villa d'Este, in the late sixteenth century. From Marie Luise Gothein, *A History of Garden Art*, vol.1, Fig. 183.
3.5. The castle and garden at Heidelberg, 1620. From Salomon de Caus, *Hortus Palatinus* (Frankfurt: De Bry, 1620).
3.6. The garden at Wilton House, Wiltshire, 1615. Isaac de Caus. From Gothein, *op. cit.*, Fig. 352.
3.7. French parterre, from the seventeenth century. Engraving by Adam Pérelle of the garden of M. de Chamlay, Paris. From Gabriel Pérelle *et al., Les Délices de Versailles et des Maisons Royales* (Paris, 1766).
3.8. The garden at Vaux-Le-Vicomte, from the chateau terrace. From Gothein, *op. cit.*, vol. 2, Fig. 391.
3.9. The gardens at Versailles, looking west along the central axis, past the pool of Latona, mother of Apollo, towards the canal. Engraving by Adam Pérelle. From Pérelle *et al., op. cit.*
3.10. The gardens of Chatsworth, Derbyshire in 1707. Leonard Knyff and Johannes Kip, *Britannia Illustrata* (London: D. Mortier, 1720).
3.11. Landscape by Claude Lorrain (1675): 'The Father of Psyche Sacrificing at the Temple of Apollo.' Anglessey Abbey, The Fairhaven Collection (The National Trust). © NTPL/John Hammond.
3.12. The Romantic reaction. Stourhead, in Wiltshire. Painting by F. Nicholson, 1813-14. Courtesy of the British Museum.
3.13. Chatsworth, transformed by Capability Brown. From F.O. Morris, *A Series of Picturesque Views of Seats of the Noblemen and Gentlemen of Great Britain and Ireland* (London: William MacKenzie, 1880).

3.14. Heveningham Hall, landscaped by Brown in 1781. Photo: *Country Life.*

Chapter 4: The Gardener as Artist

4.1. The Italianate Garden at Eaton Hall, Cheshire in 1837. From E. Adveno Brooke, *The Gardens of England* (London, 1857).
4.2. Edinburgh floral clock, 1910. Postcard. Photo courtesy of Peter Stubbs.
4.3. John Loudon's design for a suburban villa. From J.C. Loudon, *The Suburban Gardener and Villa Companion* (London: Longman, 1838), p.518, Fig. 195.
4.4. Red section of Gertrude Jekyll's flower border, *c.*1912. Photo: *Country Life.*
4.5. The blue-grey flower border at Munstead. Painting by Helen Allingham. Courtesy of Judith B. Tankard.
4.6. The white garden at Hidcote. Photograph by Peter Pritchard.
4.7. A Robinsonian wild garden in spring. Photo by Andrew Lawson.
4.8. The wild garden at Heaselands, Sussex. Photo by Tony Evans.

Chapter 5: Gardening as a Sacred Art

5.1. The lily pond, Giverny. Photo by Vivian Russell.
5.2. One of Monet's studies of the Japanese bridge, 1899. Philadelphia Museum of Art.
5.3. William Blake encounters the spirit Ololon. From *Milton*, Plate 36. Courtesy of the British Museum.
5.4. A garden in North Oxford. Photo by Louanne Richards.

Bibliography

Alan of Lille, *The Plaint of Nature*, translated with a commentary by James J. Sheridan, Pontifical Institue of Mediaeval Studies, Toronto 1980.

Alberti, Leon Battista, *On the Art of Building in Ten Books*, MIT Press, Cambridge, MA 1988.

Allan, Mea, *William Robinson, 1838–1935: Father of the English Flower Garden,* Faber and Faber, London 1982.

Allen, Paul M., *A Christian Rosenkreutz Anthology*, Rudolf Steiner Publications, New York 1968.

Barfield, Owen, *Romanticism Comes of Age*, Sophia Perennis, San Rafael, CA 1966.

—, *What Coleridge Thought*, Sophia Perennis, San Rafael, CA 1971.

Baring, Anne and Jules Cashford, *The Myth of the Goddess: Evolution of an Image*, Penguin, London 1991.

Berrall, Julia S., *The Garden: An Illustrated History*, Penguin, Harmondsworth 1978.

Bisgrove, Richard, *William Robinson: The Wild Gardener*, Frances Lincoln, London 2008.

Blake, William, *Milton*, Edited with a commentary by K.P. Easson and R.R. Easson, Thames and Hudson, London 1979.

Bockemühl, Jochen, *Dying Forests: A Crisis in Consciousness*, Hawthorn Press, Stroud 1986.

Brookes, John, *Gardens of Paradise*, Weidenfeld and Nicholson, London 1987.

Buber, Martin, *I and Thou*, translated by Walter Kaufmann, T. & T. Clark, Edinburgh 1970.

Chadwick, Alan, *Performance in the Garden*, Logosophia Press, Mars Hill, NC 2007.

Cicero, *The Nature of the Gods*, translated by Horace C.P. McGregor, Penguin, Harmondsworth 1972.

Clark, Emma, *Underneath Which Rivers Flow: The Symbolism of the Islamic Garden*, The Prince of Wales's Institute of Architecture, London 1996.

—, *The Art of the Islamic Garden*, Crowood Press, Marlborough 2004.

Clifford, Joan, *Capability Brown*, Shire Publications, Aylesbury 1974.

Coleridge, S. T. *Biographia Literaria*, 2 vols, edited by J. Shawcross, Oxford University Press, Oxford 1962.

Collingwood, R. G., *The Idea of Nature*, Oxford University Press, Oxford 1945.

Copleston, Frederick, *A History of Philosophy*, 9 vols., Image Books, New York 1962–1977.

Deely, John N., *The Four Ages of Understanding*, University of Toronto Press, Toronto 2001.

Descartes, René, *Discourse on Method*. In Descartes, *Philosophical Writings*, edited by Elizabeth Anscombe and Peter Thomas Geach, Thomas Nelson, Sunbury-on-Thames 1970.

Douglas, William Lake, 'A Garden Progress' in William Lake Douglas, *et al., Garden Design*, MacDonald, London 1984.

Farrar, Linda, *Ancient Roman Gardens*, Budding Books, Stroud 2000.

Fludd, Robert, *Summum Bonum*, in Paul M. Allen, ed., *A Christian Rosenkreutz Anthology*, Rudolf Steiner Publications, New York 1968.

Gebser, Jean, *The Ever-Present Origin*, Ohio University Press, Athens 1985.

Gervase of Tilbury, *Otia Imperialia*, edited and translated by S.E. Banks and J.W. Binns, Oxford University Press, Oxford 2002.

Giraldus Cambrensis, *Itinerarium Cambriae*, translated by R.C. Hoare, in *The Historical Works of Giraldus Cambrensis*, London 1863.

Godwin, Joscelyn, *Mystery Religions in the Ancient World*, Thames and Hudson, London 1981.

Gothein, Marie Luise Schroeter, *A History of Garden Art*, 2 vols., J.M. Dent and Sons, London 1928.

Hadfield, Miles, *A History of British Gardening*, Penguin, Harmondsworth 1985.

Harvey, John, *Mediaeval Gardens*, B.T. Batsford, London 1981.

Hobhouse, Penelope, ed. *Gertrude Jekyll on Gardening: An Anthology*, William Collins, London 1983.

—, *The Story of the Garden*, Dorling Kindersley, London 2002.

Homer, *Homeric Hymns, Homeric Apocrypha, Lives of Homer*, edited and translated by Martin L. West, Loeb Classical Library, Harvard University Press, London 2003.

Homer, *Iliad*, translated by Richmond Lattimore, University of Chicago Press, Chicago 1961.

Hughes, J. Donald, *Ecology in Ancient Civilizations*, University of New Mexico Press, Albuquerque 1975.

Hyams, Edward, *A History of Gardens and Gardening*, J.M. Dent, London 1971.

Jekyll, Gertrude, *On Gardening*, Studio Vista, London 1964.

—, *Gertrude Jekyll on Gardening: An Anthology*, edited by Penelope Hobhouse, MacMillan, London 1983.

—, *Colour Schemes for the Flower Garden*, Frances Lincoln, London 1988.

—, 'Colour in the Flower Garden' in William Robinson, *The English Flower Garden*, Bloomsbury, London 1988.

King, Ronald, *The Quest for Paradise: A History of the World's Gardens*, Whittet Books, Weybridge 1979.

Landsberg, Sylvia, *The Medieval Garden*, British Museum Press, London 1995.

Lewis, C. S., *The Discarded Image*, Cambridge University Press, Cambridge 1971.

McLean, Teresa, *Medieval English Gardens*, Barrie and Jenkins, London 1989.

Map, Walter, *De Nugis Curialium: Courtiers' Trifles*, edited and translated by M. R. James, revised by C.N.L. Brook and R.A.B. Mynors, Clarendon Press, Oxford 1983.

Merchant, Carolyn, *The Death of Nature*, Wildwood House, London 1980.

Morris, F.O., *A Series of Picturesque Views of Seats of the Noblemen and Gentlemen of Great Britain and Ireland*, William MacKenzie, London 1880.

Naydler, Jeremy, *Temple of the Cosmos: The Ancient Egyptian Experience of the Sacred*, Inner Traditions International, Rochester, Vt 1996.

—, *Goethe on Science*, Floris Books, Edinburgh 1996.

Ogilvie, R.M., *The Romans and their Gods*, Chatto and Windus, London 1974.

Plutarch, *Moral Essays*, translated by Rex Warner, Penguin, Harmondsworth 1971.

Quenot, Michel, *The Icon: Window on the Kingdom*, St Vladimir's Press, New York 2002.

Quest-Ritson, Charles, *The English Garden: A Social History*, Penguin, London 2001.

Richter, Gisela, *A Handbook of Greek Art*, Phaidon, London 1974.

Robinson, William, *The English Flower Garden*, Bloomsbury, London 1998.

—, *The Wild Garden*, Century Hutchinson, London 1986.

Sandars, N.K., *The Epic of Gilgamesh*, Penguin, Harmondsworth 1973.

Seneca, Lucius Annaeus, *Moral Epistles*, 3 vols., translated by Richard M. Gummere, Loeb Classical Library, Harvard University Press, Cambridge, Mass 1917–25.

Silverstein, Theodore, 'The Fabulous Cosmogony of Bernardus Silvestris', *Modern Philology*, 46, University of Chicago Press, Chicago 1948–9.

Silvestris, Bernardus, *The Cosmographia of Bernardus Silvestris* translated by Winthrop Wetherbee, Columbia University Press, New York 1990.

Spate, Virginia, *Claude Monet: The Colour of Time*, Thames and Hudson, London 1992.

Spinoza, Baruch, *Ethics*, J. M. Dent, London 1934.

Strong, Roy, *The Renaissance Garden in England*, Thames and Hudson, London 1979.

Telesko, Werner, *The Wisdom of Nature: The Healing Powers and Symbolism of Plants and Animals in the Middle Ages*, Prestel, London 2001.

Thompson, Ian, *The Sun King's Garden*, Bloomsbury, London 2006.

Trendelenburg, Adolf, 'A Contribution to the History of the Word "Person"', *The Monist*, XX, pp.336–363 1910.

Tucker, Paul Hayes, *Claude Monet: Life and Art*, Yale University Press, New York 1995.

Watts, Alan W., *Myth and Ritual in Christianity*, Beacon Press, Boston 1968.

Weijers, Olga, 'Contribution à l'histoire des termes *natura naturans* et *natura naturata* jusqu'à Spinoza', *Vivarium*, 16.1, pp.70–80, E.J. Brill , Leiden 1978.

Whately, Thomas, *Observations on Modern Gardening*, London 1770.

Wilkinson, Richard H., *The Complete Temples of Ancient Egypt*, Thames and Hudson, London 2000.

Wilson, Adrian, and Joyce Lancaster Wilson, *A Medieval Mirror: Speculum Humanae Savationis 1324–1500*, University of California Press, Berkeley 1984.

Yates, Frances A., *The Rosicrucian Enlightenment*, Paladin, St Albans 1975.

Notes

Introduction

1. Alan Chadwick, *Performance in the Garden* (Mars Hill, NC: Logosophia Press, 2007), p.30. Italics added.

Chapter One: The Garden in Antiquity

1. Richard H. Wilkinson, *The Complete Temples of Ancient Egypt* (London: Thames and Hudson, 2000), p.72.
2. *Epic of Gilgamesh*, Tablet IX.47ff. Translated in N.K. Sandars, *The Epic of Gilgamesh* (Harmondsworth: Penguin, 1973), p.100. Although the majority of the tablets are of the first millennium, some fragments belong to a much earlier period.
3. For a discussion of the ancient Egyptian concept of the First Time, see Jeremy Naydler, *Temple of the Cosmos: The Ancient Egyptian Experience of the Sacred* (Rochester, Vt.: Inner Traditions International, 1996), pp.91–97.
4. The lake or pool always seems to be contained within a proportional rectangle: in Fig. 1.2 it is a Golden Mean rectangle; in Fig. 1.4, a $\sqrt{3}$ rectangle; in Fig. 1.5 it is a double square ($\sqrt{4}$); and in Fig. 1.6 it is a $\sqrt{3}$ rectangle. As these proportions are unlikely to have been hit upon by chance, it is reasonable to suppose that a symbolic value may well have been assigned to each.
5. The regeneration symbolism of the lotus can be found in *The Book of the Dead*, chap. 81, and also chap. 174, which derives from the much older *Pyramid Texts*, §266.
6. Quoted in Julia S. Berrall, *The Garden: An Illustrated History* (Harmondsworth: Penguin, 1978), p.16.
7. Homer, *Iliad*, XX.7–9. trans. Richmond Lattimore (Chicago: University of Chicago Press, 1961), p.404. Compare *Odyssey*, X.350, where the nymphs are described as 'the daughters of the springs and groves and sacred sea-flowing rivers'.
8. Homeric Hymn to Pan, 19.1–3 and 19–20, in Martin L. West, ed., *Homeric Hymns, Homeric Apocrypha, Lives of Homer*. Loeb Classical Library. (London: Harvard University Press, 2003), pp.198–201.
9. Homeric Hymn to Aphrodite, 5.265ff, in West, *op. cit.*, pp.178–181.
10. Martin Buber, *I and Thou*, translated by Walter Kaufmann (Edinburgh; T. Clark, 1970), p.73.
11. William Lake Douglas, 'A Garden Progress' in William Lake Douglas, *et al.*, *Garden Design* (London: MacDonald, 1984), p.19.
12. J. Donald Hughes, *Ecology in Ancient Civilizations* (Albuquerque: University of New Mexico Press, 1975), p.50: 'Probably the Greeks first worshipped in groves and only later built temples within them; it is certain that the association of the two was so deep that temples were always set in groves of trees...'

13. Marie Luise Schroeter Gothein, *A History of Garden Art*, vol. 1 (London: J.M. Dent and Sons, 1928), pp.54–56.

14. Ronald King, *The Quest for Paradise: A History of the World's Gardens* (Weybridge: Whittet Books, 1979), p.28.

15. According to Gothein, *op. cit.*, p.71, the garden of Aristotle's successor, Theophrastus, contained not only a sanctuary to the Muses, but also a hall for teaching, and in his will the philosopher left instructions to build a second hall in the garden.

16. Gothein, *op. cit.*, p.85f.

17. *De Natura Deorum*, 3.23–28, translated in Cicero, *The Nature of the Gods*, trans Horace C.P. McGregor (Harmondsworth: Penguin, 1972), p.202f. Cicero's treatise was written in dialogue form, but the main thrust of the argument is to deny divine involvement in nature. The first philosopher to write a treatise specifically on the nature of the gods was Epicurus at the beginning of the third century BC. Many others, including his own follower Philodemus and the Stoic philosopher Chrysippus of Tarsus (both third century BC) continued the tradition, so by Cicero's time there was nothing new in subjecting the gods to rational enquiry. The process terminates in the fifth century AD in Proclus's *Elements of Theology*, in which the gods are reduced to abstractions within a metaphysical system.

18. R.M. Ogilvie, *The Romans and their Gods* (London: Chatto and Windus, 1974), p.13f.

19. *The Decline of the Oracles*, §15, in Plutarch, *Moral Essays*, trans. Rex Warner (Harmondsworth: Penguin, 1971), p.52.

20. *Ibid.*, §17, p.53f.

21. Perspective painting had been pioneered in ancient Greece by Agatharchos and Apollodoros during the latter part of the fifth century BC, but this was not applied to the depiction of the landscape until the Hellenistic period (i.e. from around the late fourth century BC). See Gisela Richter, *A Handbook of Greek Art* (London: Phaidon, 1974), p.283. The origins of the 'perspectival consciousness' that increasingly characterised the Roman relationship to nature must be traced to this period.

22. Linda Farrar, *Ancient Roman Gardens* (Stroud: Budding Books, 2000), p.144.

23. Lucius Annaeus Seneca. *Moral Epistles*, 3 vols., translated by Richard M. Gummere. The Loeb Classical Library. (Cambridge, Mass.: Harvard University Press, 1917–25), vol. 2, Ep.89, pp.391–393.

24. Statius, *Silvae*, 2.2.33ff., quoted in Farrar, *op. cit.*, p.50.

25. Statius *Silvae*, 3.1.82f. referred to in Farrar, *op. cit.*, p.59.

26. Farrar, *op. cit.*, pp.42–44 and p.86f.

27. Joscelyn Godwin, *Mystery Religions in the Ancient World* (London: Thames and Hudson, 1981), p.7.

28. Pliny the Elder, *Naturalis Historiae*, XVI, 60, 140, discussed in Farrar, *op. cit.*, p.141.

29. Pliny the Younger, Letter to Domitius Apolinarius, *Epistles*, 5.2, discussed in Berrall, *op. cit.*, p.42.

30. Adolf Trendelenburg, 'A Contribution to the History of the Word "Person"', *The Monist* XX (1910), p.356.

Chapter Two: The Garden in the Middle Ages

1. John Brookes, *Gardens of Paradise* (London: Weidenfeld and Nicholson, 1987), pp.17–24.

2. Emma Clark, *The Art of the Islamic Garden* (Marlborough: Crowood Press, 2004), p.24: 'It is in the nature of paradise to be hidden and secret, since it corresponds to the interior world, the innermost soul – *al-jannah* meaning "concealment" as well as garden...'

3. The earliest Islamic gardens can be traced to Samarra in Persia, in the middle of the ninth century. See Brookes, *op. cit.*, p.35.

4. Emma Clark, *Underneath Which Rivers Flow: The Symbolism of the Islamic Garden* (London: The Prince of Wales's Institute of Architecture, 1996), p.15.

5. For a systematic study of the religious symbolism and practical aspects of the Islamic garden, see Clark, *The Art of the Islamic Garden*, chapters 1 and 2.

6. Clark, *The Art of the Islamic Garden*, p.70.

7. *Speculum Humanae Salvationis* (from a printed edition, *c.*1475), chap. 3. Translation by Bronac Holden.

8. For example, in the late twelfth century, Walter Map, *De Nugis Curialium* ('On the trifles of Courtiers'), 2.11ff; Giraldus Cambrensis, *Itinerarium Cambriae* ('Itinerary Through Wales'), translated by R.C. Hoare, *The Historical Works of Giraldus Cambrensis* (London: 1863), p.390. In the early thirteenth century, Gervase of Tilbury, *Otia Imperialia* ('Recreation for an Emperor'), edited and translated by S.E. Banks and J.W. Binns (Oxford University Press, 2002), pp.664–677.

9. Capella, *De Nuptiis Mercurii et Philologiae*, 2.167, refers to 'dancing companies of *Longaevi* who haunt woods, glades, and groves, and lakes and springs and brooks, whose names are Pans, Fauns ... Satyrs, Silvans, Nymphs ...' Quoted in C.S. Lewis, *The Discarded Image* (Cambridge: Cambridge University Press, 1971), p.122.

10. Bernardus Silvestris, *Cosmographia*, 2.7, in *The Cosmographia of Bernardus Silvestris*, translated by Winthrop Wetherbee (New York: Columbia University Press, 1990), p.108.

11. Alain de Lille, *De Planctu Naturae*, chapter 2, translated with a commentary by James J. Sheridan in Alan of Lille, *The Plaint of Nature* (Toronto: Pontifical Institue of Mediaeval Studies, 1980), pp.73–105.

12. The distinction is first found in Michael Scot, *Liber introductorius* (1228), as well as in early 13th century Latin translations of the commentary of Ibn Rushd (Averroes) on Aristotle's *Physics*, II.1.2. See John N. Deely, *The Four Ages of Understanding* (Toronto: University of Toronto Press, 2001), pp.138–140. It is subsequently used by Bonaventure, Aquinas, and many others throughout the medieval period. For a brief history of the terms, see Olga Weijers, 'Contribution à l'histoire des termes *natura naturans* et *natura naturata* jusqu'à Spinoza,' *Vivarium* 16.1 (1978), pp.70–80. See also Chapter 5, note 7 below.

13. For types of medieval garden, see Sylvia Landsberg, *The Medieval Garden* (London: British Museum Press, 1995), chapter 1.

14. Albertus Magnus, *De Vegetabilis et Plantis* (*c.*1260), quoted in John Harvey, *Mediaeval Gardens* (London: B.T. Batsford, 1981), p.6.

15. In *The Life of St Liutgart of Wittigen*, we read: 'A delightful garden should have violets and red roses, lilies, fruit trees, green grass and a running stream.' Quoted in Harvey, *op. cit.*, p.60.

16. Teresa McLean, *Medieval English Gardens* (London: Barrie and Jenkins, 1989), p.159.

17. This image, from an early fifteenth century manuscript of the *Speculum Humanae Salvationis*, depicts the semi-mythical Assyrian Queen Semiramis, whose empire extended over Persia, and to whom the 'Hanging Gardens of Babylon' were attributed. In the *Speculum*, the queen is seen as a prefiguration of the Virgin Mary. See Adrian Wilson and Joyce Lancaster Wilson, *A Medieval Mirror: Speculum Humanae Savationis 1324–1500* (Berkeley: University of California Press, 1984), p.151.

18. Anne Baring and Jules Cashford, *The Myth of the Goddess: Evolution of an Image* (London: Penguin, 1991), pp.574–582.

19. *Ibid.*, p.575.

20. Alan W. Watts, *Myth and Ritual in Christianity* (Boston: Beacon Press, 1968), pp.101–114. According to Watts (p.101), the symbolism of the Virgin as *Rosa Mundi* is 'of the created order, *Maya*, which flowers from its divine centre.'

21. McLean, *op. cit.*, p.129. See also Theodore Silverstein, 'The Fabulous Cosmogony of Bernardus Silvestris,' *Modern Philology*, 46 (1948-9), p.105f.

22. Berrall, *op. cit.*, p.96. See also McLean, *op. cit.*, chapter 5; and Werner Telesko, *The Wisdom of Nature: The Healing Powers and Symbolism of Plants and Animals in the Middle Ages* (London: Prestel, 2001), pp.38–42.

23. Thus the late twelfth/early thirteenth century theologian and scholar, Alexander of Neckham, could write: "The stalk of the lily, when it is green, produces a most splendid flower, which changes from green to white. So must we persevere in the best of behaviour so that, immature plants as we are, we may attain to the whiteness of innocence. Furthermore, the whiteness of this flower does not turn back into green, and even so our continence must not succumb to the enticements of this dissolute life." Quoted in McLean, *op. cit.*, p.162.

24. Harvey, *op. cit.*, p.*x.*, warns us that in order to understand medieval horticulture, 'it is necessary to accept the existence of a completely different set of values from those of the modern world in which we live. Fundamentally the Middle Ages exemplified a human society based upon transcendental and spiritual motives ...'

Chapter Three: From the Renaissance to the Eighteenth Century

1. Jean Gebser, *The Ever-Present Origin* (Athens: Ohio University Press, 1985), p.16, remarks on 'the unprecedented inner struggle that occurred in artists of that generation of the fifteenth century during their attempts to master space' and of 'the artist's inner compulsion to render space – which is only incompletely grasped and only gradually emerges

out of his soul toward awareness and clear objectification – and his tenacity in the face of this problem because, however dimly, he has already perceived space'.

2. For the emergence of perspectival consciousness, see Gebser, *op. cit.*, pp.11–23.

3. Carolyn Merchant, *The Death of Nature* (London: Wildwood House, 1980), chapters 7–9.

4. Leon Battista Alberti, *De Re Aedificatoria* (1443–1452) translated as *On the Art of Building in Ten Books* (Cambridge, Mas: MIT Press, 1988).

5. Roy Strong, *The Renaissance Garden in England* (London: Thames and Hudson, 1979), p.15.

6. *Ibid.*, p.21.

7. Frances A. Yates, *The Rosicrucian Enlightenment* (St Albans: Paladin Frogmore, 1975), p.38f.

8. *Ibid.*, p.100.

9. *Ibid.*, p.101.

10. Robert Fludd, *Summum Bonum* (1629), for example, described the Rosicrucian work as a 'mediation' through which 'the earthly has been opened to the entry of the joys of Paradise ...' See Paul M. Allen, *A Christian Rosenkreutz Anthology* (New York: Rudolf Steiner Publications, 1968), p.374.

11. Strong, *op cit.*, pp.147–164.

12. King, *op. cit.*, p.155; Berrall, *op. cit.*, p.198.

13. According to Penelope Hobhouse, *The Story of the Garden* (London: Dorling Kindersley, 2002), p.152, Le Nôtre studied Descartes. However, it seems more likely that his main source was Jacques de Boyceau's *Traité du Jardinage* (*Treatise on Gardening According to the Reasons of Nature and of Art*), published posthumously in 1638, which set forth the principles of formal garden design, emphasising the importance of geometry, architecture and arithmetic. See Ian Thompson, *The Sun King's Garden* (London: Bloomsbury, 2006), p.94 and p.30f. The approach of Boyceau should not be confused with that taken in the Islamic garden, in which geometrical principles were employed to make the earthly garden into a mirror of the heavenly paradise. In the European gardens of the sixteenth and seventeenth centuries, (with the possible exception of gardens such as Heidelberg and Wilton) geometry was employed either to create meaningless decorative patterns or as an aid to the creation of perspectival illusions, which Le Nôtre especially excelled at.

14. King, *op. cit.*, p.157.

15. According to Thompson, *op. cit.*, p.230, by 1672 the number of fountains had reached the incredible total of 2,456, which meant that there was a constant problem of supplying the garden with sufficient water to keep them all functioning.

16. Berrall, *op. cit.*, p.200f.

17. Hobhouse, *op. cit.*, p.153.

18. Thompson, *op. cit.*, pp.253–256.

19. Duc de Saint -Simon, quoted in Thompson, *op. cit.*, p.4.

20. Thompson, *op. cit.*, p.258.

21. René Descartes, *Discourse on Method*, Part 6 in *Philosophical Writings*, edited by Elizabeth Anscombe and Peter Thomas Geach (Sunbury-on-Thames: Thomas Nelson, 1970), p.46.

22. Quoted in King, *op. cit.*, p.179.

23. Gothein, *op. cit.*, vol 2, pp.281–288.

24. Thomas Whately, *Observations on Modern Gardening* (London, 1770), p.1.

25. Charles Quest-Ritson, *The English Garden: A Social History* (London: Penguin, 2001), p.120.

26. Joan Clifford, *Capability Brown* (Aylesbury: Shire Publications, 1974), p.43.
27. Quest-Ritson, *op. cit.*, p.145.
28. Thompson, *op. cit.*, pp.82–84.

Chapter Four: The Gardener as Artist

1. Owen Barfield, *Romanticism Comes of Age* (San Rafael, CA: Sophia Perennis, 1966), p.234.
2. S.T. Coleridge, *Biographia Literaria*, ed. J. Shawcross (Oxford: Oxford University Press, 1962), vol. 2, p.257.
3. John Loudon, *The Suburban Gardener and Villa Companion* (1838), quoted in Miles Hadfield, *A History of British Gardening* (Harmondsworth: Penguin, 1985), p.258.
4. Gertrude Jekyll, 'Colour in the Flower Garden' in William Robinson, *The English Flower Garden* (London: Bloomsbury, 1998), p.121.
5. Gertrude Jekyll, *Colour Schemes for the Flower Garden* (London: Frances Lincoln, 1988), p.71.
6. Gertrude Jekyll, *On Gardening* (London: Studio Vista, 1964), p.14.
7. Something similar can be experienced at Sissinghurst, another garden influenced by Jekyll's approach.
8. Penelope Hobhouse, *Gertrude Jekyll on Gardening: An Anthology* (London: MacMillan, 1983), p.21.
9. Gertrude Jekyll, *On Gardening*, p.12.
10. In this respect, Jekyll's approach is strongly reminiscent of that of Goethe, for whom the intensification of the act of sensory perception to the point at which it becomes insight (*Anschauung*) was one of the keys to the redemption of our relationship with nature. See Jeremy Naydler, *Goethe on Science* (Edinburgh: Floris Books, 1996), p.50.
11. Jekyll, *On Gardening*, p.12.
12. Jekyll, *Wood and Garden*, in Hobhouse, *Gertrude Jekyll on Gardening*, p.23.
13. Gertrude Jekyll, *The Garden*, September, 1899. Quoted in Richard Bisgrove, *William Robinson: The Wild Gardener* (London: Frances Lincoln, 2008), p.190.
14. William Robinson, *The English Flower Garden* (London: Bloomsbury, 1998), p.120: 'Nature is a good colourist, and if we trust her guidance we never find wrong colour in wood, meadow, or on mountain.' For Nature as the supreme artist, see *ibid.*, p.29.
15. William Robinson, quoted in Mea Allan, *William Robinson, 1838-1935: Father of the English Flower Garden* (London: Faber and Faber, 1982), p.133.
16. Robinson, *The English Flower Garden*, p.86, where he quotes the first lines of Shelley's famous translation of the Homeric Hymn, 30, 'To Earth, Mother of All':
    > O universal Mother,
    > who dost keep
    > From everlasting thy
    > foundation deep,
    > Eldest of things, Great
    > Earth, I sing of thee.
17. Robinson, *The Wild Garden* (London: Century Hutchinson, 1986), p.14.
18. Robinson, *The English Flower Garden*, p.159
19. *Ibid.*, p.*viii*.

Chapter Five: Gardening as a Sacred Art

1. Paul Hayes Tucker, *Claude Monet: Life and Art* (New York: Yale University Press, 1995), p.175.

2. Monet, quoted in Tucker, *op. cit.*, p.178.

3. Quoted in Virginia Spate, *Claude Monet: The Colour of Time* (London: Thames and Hudson, 1992), p.216.

4. Mirbeau, quoted in Spate, *op. cit.*, p.169. See also p.200. Translation adapted.

5. Monet, in a letter to Gustave Geffroy, quoted in Spate, *op. cit.*, p.205.

6. Michel Quenot, *The Icon: Window on the Kingdom* (New York: St Vladimir's Seminary Press, 2002), p.87.

7. Michael Scot, writing in the early thirteenth century, stated in his book *Liber introductorius*: 'God is Nature naturing (*Natura naturans*) and therefore overcomes Nature natured (*Natura naturata*).' [Deus sit natura naturans et ideo superet naturam naturatam.] Quoted in Weijers, *op. cit.*, p.71. It is in the conception of *Natura naturans* that the world of nature spirits is accommodated.

8. Baruch Spinoza, *Ethics* (London: J.M. Dent, 1934), Part 1, proposition 29, p.24. See also Weijers, *op. cit.*, pp.77–80.

9. S.T. Coleridge, *Biographia Literaria*, vol. 2, p.257. See also Owen Barfield, *What Coleridge Thought* (San Rafael, CA: Sophia Perennis, 1971), pp.22–25, and p.199.

10. For Goethe, too, it is precisely through intensifying our awareness of 'eternally creative nature' that we enable ourselves to become spiritual participants in nature's creative process. See Naydler, *Goethe on Science*, p.101.

11. William Blake, *Milton*, edited with a commentary by K.P. Easson and R.R. Easson (London: Thames and Hudson, 1979), Part 2, Plate 36, p.121.

12. This was well understood by the Goethean scientist, Jochen Bockemühl, whose book *Dying Forests: A Crisis in Consciousness* (Stroud: Hawthorn Press, 1986), is an invaluable source of insight. See especially, *Dying Forests*, p.30.

13. Octave Mirbeau, quoted in Virginia Spate, *op. cit.*, p.216.

14. Robinson, *The English Flower Garden*, p.159.

Index